Letters from John Dewey/Letters from Huck Finn: A Look at Math Education from the Inside

by Barry Garelick

Modern Educator Press

Dedication

To my Mom who started teaching later in life, to my wife Ginny and my daughter Angela who kept me from quitting, and to all my students.

The letters from John Dewey originally appeared in the blog Edspresso in slightly different form. The letters from Huck Finn (with the exception of Chapter 10) originally appeared in the blog Out in Left Field, also in slightly different form.

Second Edition

ISBN-13: 978-0692509562
ISBN-10: 0692509569

Library of Congress Control Number: 2015913031

Table of Contents

Preface to the Second Edition

Those who are new to the disagreements these days on how best to teach math, may know only of the Common Core Math Standards. But before there was Common Core, one of the main sticking points in arguments about math education was a set of standards written by the National Council of Teachers of Mathematics (NCTM). Originally written in 1989, and then revised in 2000, they were criticized and blamed for the restructuring of math classes to make them less computational, less "rote learning", less "learning concepts in isolation" and other mischaracterizations of traditional math—with one of the main goals being that students learn to "think like mathematicians".

When I wrote the series of letters under the pseudonym John Dewey, the NCTM standards were going strong and therefore many of the letters mentioned them. But the focus these days is on the Common Core Math Standards which have been adopted by 45 states. And though these standards have some good features (like teaching the standard algorithms and calling for fluency of basic addition/subtraction, multiplication/division facts), there are enough code words (like "students shall explain" and "students shall understand") that these standards have leant themselves to interpretations similar to the old NCTM standards.

The second half of the book consists of letters/episodes written under the name Huck Finn, and Common Core starts to surface more. This book spans a six year period from 2006 to 2012 and thus provides a sense of history of how things have progressed when NCTM was the watchword among math reformers, to a time when the torch was passed to the bearers of the Common Core. In short, not much changed except the names of things.

Speaking of names, readers of the first edition have often asked me why I took the name John Dewey for the first set of letters. I could come up with fancy sounding reasons but the short answer is that what I talk about and strive for in education is almost the complete opposite of Dewey's vision of "teaching the whole child". That vision was instrumental in begetting today's student-centered classrooms, in which any real and effective teaching gives way to "formative assessment" and "facilitation".

Another reason for the name John Dewey is that the set of letters describe what was going on in two Ed. School classes

I was taking, and thus represent the theoretical (or Dewey-like) aspect of teaching. By contrast, the second half of the book takes place in real classrooms, so the disconnect between the theory and actual practice was much more evident. To that end, Huck Finn seemed an appropriate name for a neophyte caught between opposing realities.

Barry Garelick
August, 2015

Introduction:

In Which I Explain Myself Without Apology

This book is a collection of letters that I wrote which chronicle my experiences in a math teaching methods class in Ed. school (using the name John Dewey) and my experiences student teaching (using the name Huck Finn). I teach mathematics in California. I have a degree in the subject and an intense interest in how it is taught.

When my daughter was in elementary school I saw things I didn't like about the way she was being taught math. I was also tutoring high school students in math and saw disturbing weaknesses in basic math skills. This caused me to embark in research about what is going on in math education. I decided that the way I could possibly make a difference was to teach mathematics in middle or high school. In the fall of 2005, with six more years left until I could retire, I enrolled in education school.

By way of a short background, the debate over how math is best taught in K-12 (and which is known as the "math wars") has been going on for many years, starting perhaps in the early part of the 20th century. The education theory at the heart of the dispute can be traced to John Dewey, an early proponent of learning through discovery. Fast forward to 1957 when Sputnik was launched and the New Math era began in earnest, which continued until the early 70's. Then came the "back to basics" movement, and in 1989 the National Council of Teachers of Mathematics (NCTM) came out with *The Curriculum and Evaluation Standards for School Mathematics,* also known as the NCTM standards.

The NCTM's view was that traditional teaching techniques were akin to "rote memorization" and that in order for students to *truly learn* mathematics, the subject must be taught "with understanding". Thus, process trumped content. Showing how students obtained the answer to a problem was more important than getting a right answer. Open-ended ill-posed problems became the order for the day. The prevailing education groupthink was (and still is) that teaching the mathematical procedures for particular types of problems was just more rote. Such approaches didn't teach students "higher order thinking skills", "critical thinking" and many other terms that are part of the education establishment's lexicon.

By the time I enrolled in Ed. school, I pretty much knew what I was in for. I was well acquainted with the theories of teaching and learning which dominated the education establishment in general and education schools in particular. Nevertheless, I was surprised at what I heard when going through the candidate interviews, which was part of the application process. Future teachers of science and math were herded in one group and given a brief talk by the coordinator of secondary education. Among her opening remarks was the announcement that "The way math and science are taught today is probably not how you were taught when you were in school." A few sentences later, the coordinator, with index finger pointing to the ceiling for emphasis, said "Inquiry-based learning!" Though a bit unnerved, I at least knew where I was.

All in all, my Ed. school experience had some redeeming features. Most of my teachers had taught in K-12, and had valuable advice about classroom management problems and some good common-sense approaches to teaching that didn't rely on nausea-inducing theories. Also, I learned how to make it sound like my approach to teaching was what was being taught. I learned to talk about discovery approaches and small group exercises—no one has to know that such techniques are not going to be your dominant teaching approach. In short, since future teachers will be working in a bureaucracy that is often dictated by the groupthink of the education establishment, Ed. school serves the purpose of teaching survival techniques.

Sometime after I took my first course, I decided to write a series of letters documenting my experience in Ed. school, using the pseudonym of John Dewey. There was a new education blog that had emerged called Edspresso, edited by a genial and talented young man named Ryan Boots. (Unfortunately, he left Edspresso several years ago). I pitched the idea to him, asking him what he thought. He responded almost immediately along the lines of "An Ed. school mole writing about his experiences? When can you start?"

My series of letters for Edspresso covered mainly one class—the beginning math teaching methods class. The letters proved to be very popular and many people left comments—some supportive, and some very angry. I wrote the letters almost in real time—there was perhaps a one or two week delay between the letter I was writing and the events of a particular class.

As I progressed through the class, I noticed that while my views on teaching may have differed from that of the teacher (an adjunct professor who I refer to as Mr. NCTM), there were certain views that we shared in common. We were both around the same age, and he had taught high school math for 30 years. He had very good advice and it was clear that he liked me. I came to the realization that though there were vast differences in teaching philosophies within the teaching profession, one had to work with fellow teachers as well as the people in power on a daily basis. The trick would be to find a situation in which I could be loyal to how I believed math should be taught, and find that common bond with the other teachers and the administration that would allow us all to get along.

I decided to stop writing the letters when the math teaching methods class ended. This was not only because of the time involved in writing them, but because of a fear that their continuation would ultimately lead someone to discover the identity of the author. I didn't want to ruin any chance of obtaining a teaching credential, nor to be blackballed from any teaching positions because of differences in teaching philosophy.

After several years, I had completed all my coursework and was ready to move on to student teaching. I had a few months to go until retirement, and then could take on the commitment for the remaining task. I felt that this phase called for a resurrection of John Dewey, but my initial draft of a letter seemed forced and the voice of Mr. Dewey no longer seemed appropriate.

Around that time, I had the good fortune to have seen a performance of Hal Holbrook as Mark Twain. Mr. Holbrook was 85, so I knew this might be my last chance to see him. The performance lived up to everything I had heard about it, but one part of the evening stood out. He did a reading from Huckleberry Finn that was extremely moving and convincing. I heard the voice of a naive young boy commenting on rather serious matters over which he had no control, but about which he was beginning to form life-changing opinions. I realized the next day that Huck Finn was the perfect choice for the author of the letters about student teaching, immersed in the polarized world of education, and drifting along the ideological, political and cultural divide.

I asked Katharine Beals who runs the blog "Out In Left Field" if she wouldn't mind publishing some letters from Huck Finn about the process of becoming a math teacher. She was excited about this and so I decided to give it a go. I was grateful for her taking Huck in; she is known as "Miss Katharine" in the letters. The name seemed to fit her quite well.

The first two Huck Finn letters are about a year apart, and then they follow the student teaching. I couldn't write those in real time since the teaching kept me rather busy, so I wrote the letters after I finished. After another year I wrote six more episodes, this time looking at Huck's experience as a substitute teacher.

I'm trying to think of something profound and moving to close with here and the best I could come up with was "For anyone wanting to make a movie based on these letters, please don't have me played by Matt Damon." Actually, a comment I received on one of the Huck Finn letters from Niki Hayes, a former teacher and principal, is much better I think, so let me close with that and offer it to you as advice:

So you learned what teaching is about: The dispensing of content information so that kids don't have to "struggle" repeatedly to understand it (which makes most humans turn off the learning switch) AND experiencing those wonderful young eyes that make you want to be a better teacher and person. You'll always remember these kids because they were your first "tutors." Let me assure you, there will be many more as you enter the special land of teaching.

Letters from John Dewey

1. By Way of Introduction

By way of introduction, my name is not John Dewey but it will have to do for now. I am in Ed. school (night classes) in a state and county that shall remain nameless as shall my state of mind. I have decided to teach high school math when I retire in five years. I majored in the subject, and have a burning desire to make sure kids learn it correctly.

I recognize that there are probably more than a few blogs out there that operate on the premise of the anonymous mole in Ed. school telling all. So what makes me different? For one, other blogs simply pale in comparison. For another, no other Ed. school mole blogs are writing about math.

Math education is in a shambles, starting from the so-called standards put out by the National Council of Mathematics (NCTM) in 1989 and revised in 2000. These standards were then copied by many states that thought they were great. State boards of education paid no mind to the shrieks of horror from mathematicians, simply not believing that the resulting standards took the math out of mathematics in the name of fun, and whose approach for eliminating the achievement gap eliminated the mastery of any math knowledge that matters. The well-intentioned but ill-conceived standards have actually widened the gaps between the rich and the poor by motivating those who can do so to hire tutors for their children, to enroll them in learning centers like Sylvan and Kumon, or to put them in private schools.

Few refuges exist from the multi-colored tomes that adhere to NCTM-based standards posing as math textbooks. No one is safe from this modern day invasion of the body snatchers. And just like in the movie, those with the power to do something have already been taken over by the seed pods of Ed. school dogma. Those who resist are told that everything they've heard is false. It's just ideology not fact, they are told, propagated by math professors who are evil seed pods themselves desiring to turn your kids into dull, listless mathematicians like them. And just to make sure you're okay with that, they are then told that under these math reform programs at least your kids will learn critical thinking skills. Oh, good. Now I feel better.

I exaggerate, just a bit. Well, no. I take it back. I don't exaggerate at all. The seed pods who infiltrated NCTM are the same ones who infiltrated the Ed. schools. As one mathematician I know puts it, the inmates are running the insane asylum. Perhaps it started with the original John Dewey, but something tells me if he saw what is going on now he would say, "No, that's not what I had in mind. And who was the idiot who came up with block scheduling?"

I am a new breed of warrior that is trying to infiltrate from the inside by actually teaching math as it should be taught. This means—I am told by teachers getting ready to retire—that I should teach in a private or a charter school.

These teachers have had enough. They view me in the same way as a well-fed restaurant patron looks at the new customers gorging themselves: "How can they eat when I'm not hungry?"

The seed pod way of thought begins early in Ed. school. For me, it began even before my first day of classes. The short-listed candidates for admission to the grad school of education had to come in for an orientation and interview. Among the many Ed. school adages we were told that day was: "The way science and math are taught today is not how you were taught." This said with a kind of taunting, challenging quality with some mumbo jumbo thrown in about "inquiry-based" learning. Another was "The textbook is a resource, not a curriculum." From what I hear, this "content doesn't matter" approach only gets worse.

I have just finished my first introductory class which though steeped in the various theories of learning that abound, did not require any essays extolling the "content doesn't matter" philosophy (though you certainly weren't discouraged from expressing such sentiment). The course I'm dreading is coming up this fall—a class in teaching methods for math. The syllabus reads like a promotional brochure for the NCTM standards.

Right now I'm studying for the Praxis II test in mathematics—required for licensure and for graduation. It is given by the Educational Testing Service (ETS), the folks that produce the SAT, GRE and hosts of other standardized tests. The Praxis II math content test covers algebra through calculus and requires the use of a graphing calculator. My advisor told me I needed one for the test. To find out where her sympathies lay, I wrote her an email with my observation that with the increasing use of calculators, the rudiments of math were becoming a lost art.

My advisor replied that there was a time when writing down the plus and minus symbols was considered a "crutch" and that what is "basic" or "fundamental" or "rudimentary" is by no means fixed and never has been. This is someone with a master's degree in math. I was about to ask her if the multiplication facts had changed, and how long had we been using the plus and minus symbols but decided against it. From what I hear, she's the one teaching the course I have to take in the fall.

No one is safe from this modern day invasion of the body snatchers. And just like in the movie, those with the power to do something have already been taken over by the seed pods of Ed. school dogma.

2. Kicking the Ed School Blues

Some people are telling me Ed. school is just a minor annoyance, and that once the door to my classroom is closed, the class is mine to do with as I wish. Others tell me pick a school where there are no "math police" who make sure I teach the edu-fad du jour.

Of all the things I've heard, two in particular stand out. One from a friend who asked if I thought I was making a difference with this little venture into blog space. The other asked whether I thought I'd be making a difference teaching in a system that prevents effective math teaching in a world infiltrated by NSF, NCTM/Ed. school dogma and math police.

I don't know the answer to the first question. But I'm in Ed. school, where there are no wrong answers. So here goes. Will this little blog venture make a difference? Well, what I do know is that Ed. schools—without benefit of blogs or internet cafés—have made a huge difference in this country. A bad one. Therefore, the more people informed of the debate the better, particularly those on the fence.

This brings up the second question: if the seed pod infiltration is so effective (see my last letter for what this metaphor means) what is the chance for change with only a few enlightened teachers battling the math police?

My answer to the second question is based on the fact that I've never had an original idea in my life. Being part of the baby boomer generation means that whatever so-called original idea is in my head is also in the heads of thousands of other people. Therefore, many people getting ready to retire and who have science or math backgrounds may also be looking into teaching. There's strength in numbers. (No pun or content intended.) If my class at Ed. school is any indication, four of the five future math teachers in the room were of my vintage. And if it's any consolation, I believe that four out of the five future math teachers tended to ignore what was taught. (I think this might extend to other disciplines as well).

In the class I just took, the professor one night espoused the ubiquitous Ed. school philosophy that one of the biggest hurdles to conquer in teaching math is students' math anxiety. He provided an example. He handed out a problem that asked in what position was a table held while moved, if it produced a scratch on the floor that was in a northeasterly direction. The problem could have many answers, a concept beloved by Ed. school types who believe that problems with only one correct answer limit students' critical thinking skills. "Open-ended" problems with many answers, on the other hand, reduce math anxiety because it relieves the pressure to produce THE correct answer. Students are thus liberated to be creative and use "higher order thinking skills". I pointed out that the problem was not so much open-ended as it was ill-posed.

"Yes, it is ill-posed," he agreed. There were no arguments in this class; only insights, discussions, and agreement. This is Ed. school: there are no wrong answers. Just the "greater truth" which will eventually prevail. No such epiphanies occurred that night, however. One student said that the scratch-on-the-floor problem actually made her more anxious because she wasn't sure what she was doing wrong. The teacher said "Yes, I agree," and concluded that perhaps the best way is to tell the students at the outset that there is more than one right answer. I suggested asking the students what additional information should be provided to make the problem well defined. "I agree," he agreed again.

He talked some more about math anxiety. The Ed. school of thought holds that if you just relax and get over the anxiety, the greater truth will prevail. Not a word about how inadequate preparation may play a role. "At-risk" students are particularly vulnerable to math anxiety according to Ed. school wisdom. One instructor the professor knew was quite good with such students. He told how she gave each student a name having to do with a concept in algebra. One student was called "perfect square trinomial"; another was "binomial", and so forth. (They may have had name tags).

Their task was to learn how each of them "related" to one another, thus forcing them to learn what these terms meant. Which would be great if the only goal of an algebra class were to master vocabulary and get in touch with one's inner polynomial. Perhaps this is all that is expected of these at-risk students, since they seem to have different "learning styles" than the rest of us.

I said nothing after he told us this tale, and there were no comments from anyone else either. In the papers I wrote for the class, I challenged many of the beliefs about how best to teach math. I don't know what the others said. I only know the professor said he agreed with me. I'm trying to enjoy this illusion while it lasts.

In the papers I wrote for the class, I challenged many of the beliefs about how best to teach math. I don't know what the others said. I only know the professor said he agreed with me. I'm trying to enjoy this illusion while it lasts.

3. Divergence and Convergence

For those of you who read my last missive regarding the highly agreeable professor, I'm sure it comes as no surprise that he would be the first to agree with the comments posted that many problems have only one answer—the right one. I didn't mean to pick on the guy so much. He was merely echoing Ed. school wisdom about math education. In Ed. school parlance, when more than one answer exists for a question, the thinking used to come up with answers is called "divergent". When only one answer exists, the thinking is called "convergent".

In Ed. school, "divergence" is considered a good thing and "convergence" looked upon with disdain. I think Ed. school teachers take an oath to uphold these beliefs as part of an attempt to turn math into a "divergent thinking" type of subject like social studies or English. Such thinking reflects a significant and depressing lack of understanding of what math is about. A math professor recently commented to me about this lack of understanding with respect to how it is taught:

> *One problem I see that arises from how math is taught*
> *before college is that we get some math majors (who don't all*

stay majors) who have a completely incorrect notion of what math is. I'm not sure what they think it is, but when they have to take a course like abstract algebra and are asked to do proofs, they think this has nothing to do with math. Exposure to proofs in high school geometry would go a long way toward correcting such misconceptions early on.

Unfortunately, the emphasis on proofs in geometry has been de-emphasized over the years, thanks in large part to the NCTM standards (see first letter in this series) and what Ed. schools think math is about. The proofs that exist in today's high school geometry courses are trivial; many textbooks have turned most theorems into postulates so that geometry has become a collection of "taken on faith" propositions with little to no proofs offered. Geometry classes have become mostly memorization of formulae (areas, volumes, surface areas of volumes).

As part of my recent Ed. school course, I was required to log in 15 hours of field experience by observing math classes at the high school level. I decided to visit a teacher who teaches honors geometry at the local high school. I had tutored one of her students a few years ago. I recall that she had supplemented the almost proof-less textbook by giving the students proofs to do, but even so, the number of proofs was far less than what previous generations had to do in non-honors courses.

I was surprised when I visited her class and saw how she "proved" the Pythagorean Theorem. (This is one theorem the textbook had not yet turned into a postulate.) She handed out sheets of paper on which were drawn a right triangle with three squares extending from each of its three sides. There is a famous proof in which the two smaller squares can be shown via congruence theorems to fit into the larger square of the hypotenuse. Her version, however, was to have the students cut out pieces of the two smaller squares by cutting along lines marked within them and assemble the resulting pieces, like a jigsaw, into the big square of the hypotenuse.

This was how she proved the theorem. (Oh, excuse me. This is how she had the *students* prove the theorem.) "Does this prove the theorem?" she asked. The students said yes, because it showed that the areas of the squares of the two legs in a right triangle equal the area of the square of the hypotenuse. Which is correct for the particular triangle on the sheet of paper she handed out.

While "proof by dissection" is a valid method, it requires some justification for the various dissections and moves. There was no discussion of why the teacher drew the lines within the two smaller squares where she did, what they represented, or how the procedure could be generalized for right triangles for any size. I asked her later if she offered any other proofs of the theorem. "No," she said. "We don't spend much time on the Pythagorean Theorem in the Honors class simply because they've learned it before."

I visited another class as part of my assignment and observed an algebra 1 lesson. The teacher of that class was actually quite good. He has a provisional license to teach and is getting his degree from the school I attend; even has the same advisor as I do it turned out. One exchange in class caught my attention. He announced they would be starting a new chapter in the book. A student asked "Is there any math in this chapter?" to which the teacher, straight-faced, replied "Yes, but not too much." Since the algebra text itself is not very good (same authors as the proof-less geometry book) I wondered perhaps if the two of them were making a statement about the book. That wasn't it. He said the student was a wise-ass. "I was just giving it back to him," he said.

I followed up with an email and asked what he thought of NCTM's standards and whether our advisor liked them. No reply. I think I may have scared him. That's too bad; I'm only trying to help. I hope he remains an excellent teacher.

4. Ed School and the Stomach Flu

It has been a while since I have written. I felt I had said enough about my previous class and wanted to wait until the start of the real thing: the Math Teaching Methods class, the first session of which I just attended.

A new class is always reason for anxiety, particularly after a semester with the agreeable professor. You are faced with someone new who has different rules and expectations—and may not be as agreeable. I am in a class in which the teacher is, shall we say, an adherent of the National Council of Teachers of Mathematics (NCTM) and its standards. In fact, the NCTM standards and our understanding of same make up a portion of the syllabus.

Our first assignment is a comparison of those standards with the math standards for the state in which we reside for a particular "content standard", grade level, and "process standard. The content standard describes what students are supposed to learn. The process standard describes how they are supposed to learn it. I got assigned Geometry/11th grade/representation.

"What is 'representation'?" I hear you asking. Expressing things in different ways, I think. You can use a graph to express a function, or a table of values, or a formula, for example. Which one is best to analyze the problem at hand, I think is what they're getting at but they go on and on in the standards, bringing in all sorts of ways to show things which might be good things to mention as an aside, but to devote so much class time to it supplants the basics that they are supposed to be learning. (And which educationists think is mundane, and mind numbing.)

I was suffering from stomach flu when I was reading NCTM's standards. My wife asked why I was pushing myself like that. My response was along the lines that as long as I was vomiting anyway, may as well take advantage of it.

In any class, there is a phenomenon of competition and sussing one another out, and trying to please the teacher. And since we all have math backgrounds and can no longer pretend that we're different or better than classmates who do not have math backgrounds, the competition is a bit more intense. Although this is quite normal and, some might say, healthy, it can also be an insidious part of a learning experience, particularly where ideologies and the NCTM standards are concerned.

Case in point: In our first class we were broken up into groups, and asked to look at a particular content standard for our state. Our group got Algebra 1. We were given 15 minutes to look them over and characterize them. "What do you mean by characterize them?" one person asked. The answer to that question was so vague I don't remember it—

something along the lines of "describe their relevance with respect to process and content" but even that's too specific to do the answer justice.

What grabbed my attention was the standard that required that students be able to solve quadratic equations in one variable with a graphing calculator as the primary tool. My feelings about graphing calculators aside, I noted to the others in my group that it said nothing about students learning the quadratic formula, much less its derivation. A woman in my group, in apparent defense of the standard, told me her daughter didn't have to learn the quadratic formula in Algebra 1. I pointed to that standard and said "You're looking at the reason why."

When our turn came to report our findings to the class, I said the Algebra 1 standards were vague and allowed teachers to not teach the quadratic formula. Some others in the room agreed. The teacher—Mr. NCTM—in a thinly veiled, poker-faced support of anything resembling NCTM standards, responded that the standards were in fact, not "prescriptive". This generated some discussion about giving teachers flexibility and I found myself in a debate with a bright young man who although agreeing that the quadratic formula should be taught was also caught in an unconscious effort to please the teacher.

He found himself arguing that the standards were what must be taught "at a minimum", that the non-prescriptive nature of the standard gave teachers flexibility to go beyond the minimum. It apparently didn't bother him that the minimum was inadequate and that some teachers—and textbooks—would not go beyond it. He argued with gusto, however, enjoying the limelight, not knowing prior to tonight's class that state standards existed, that NCTM existed, and that NCTM standards even existed. All he knew was some standards are better than none and that he was pleasing Mr. NCTM, and wanted to be right about something of which he knew very little. It did not bother him that the class was concerned with teaching to the NCTM standards and their look-alikes.

Later that week I worked on my assignment—comparing my state's standard and NCTM for geometry. NCTM's standard, in part, says: "Students should see the power of deductive proof in establishing the validity of general results from given conductions." My state's standards said in entirety (emphasis added): "A gradual development of formal proof is *encouraged*. Inductive and intuitive approaches to proof as well as deductive axiomatic methods *should* be used." After reading those, I was hit with the stomach flu, I think.

5. The NCTM Focal Points and an Interesting Video

Many thanks to all of you who wished me speedy recovery from my stomach flu. I assure you I am fine. A momentary lapse and now I'm all the stronger for it. In the meantime, the world has become a more interesting place, what with NCTM announcing its new "focal points", and The Wall Street Journal and The New York Times reporting that NCTM has at long last come to its senses and is going to a "back to basics" approach in its math standards. NCTM vigorously denies that the focal points represent such change and has stated that the focal points are merely a continuation of the 1989 standards, which in the view of many parents, mathematicians and (shh) teachers, ensured that computational algorithms, manipulation of symbolic expressions, and paper and pencil drill did not play a leading role in any school's math curriculum.

Both New York Times and Wall Street Journal suggested that perhaps the focal points represented the approach that Singapore has used in its math program which has been held responsible for propelling that island nation's 4th and 8th graders to the number 1 spot in international math testing for the last decade. Singapore does not apologize for having students memorize the number facts.

NCTM, however, finds the word "memorization" offensive which may explain the wording of one of the "focal points" for second grade: "Children use their understanding of addition to develop quick recall of basic addition facts and related subtraction facts." There. Every addition and subtraction fact must be understood before committing to—forgive the word—memory. For the folks at NCTM, admitting that memorization even exists is like Thomas touching the wounds of the great Savior. So maybe NCTM has a point when they vigorously deny that the focal points represent a change, and say this is the way they've always done things. Time will tell; if states change their standards and school districts start dropping Everyday Math, Investigations and other atrocities and start adopting Singapore Math, or Saxon Math, we'll know something happened.

Meanwhile, here in Ed. school, all is well. Our instructor, Mr. NCTM, assured us the focal points were just a clarification, and that nothing was different. Then we set about watching some videos of teachers using the discovery method in class. One video showed a teacher engaging her students in an activity in order to teach them about slopes. I tell you, these kids were busy. She had them measuring the volumes of two mystery liquids, weighing them, filling out a chart with the values, computing the ratios of mass to volume, and all the time, she asked questions. They plotted the ratios of mass to volume of liquids and obtained two slopes, checked their results with a graphing calculator, and she questioned them about what the graphs told them about the liquids. Luckily we saw that video on two occasions, because the first time I lost track of what these kids were supposed to be doing. I almost lost track the second time, but it finally sunk in. "Oh, she's teaching them how to interpret what the slope represents." I came to the following conclusion about her technique: If I had had a class like that in school, I would have grown up hating math.

Another video showed a teacher with his students standing around a table in the center of the room while he explained that day's assignment. (We were to take note of the fact that this standing-around-the-table approach was very effective because it ensured that all students were attentive and no one was asleep at their desk. And, of course, you can do this if you have a class with only 12 students as was the case in this video.)

This lesson was about parabolas, how the various constants in the vertex form of the equation for a parabola governed its shape, location and direction. He had them split into four groups, each group exploring what happens when you vary one particular constant. They were to use colored pipe cleaners to show the various parabolas on a poster. When through, the students all convened around the central table again and the teacher asked many questions which the students answered, some correctly, some not. There was no "That's right, that's wrong", just more questions.

The teachers in both videos were extremely good at what they were doing, which brought home an unsettling realization to me: You can be very good at doing something that is absolutely horrible. And when you see teachers like these who are very good at what they do, if you didn't know any better you would try to emulate having students spend an entire class period bending pipe cleaners into parabolas and gluing them on poster boards.

After the videos we voiced our observations to the teacher. One woman said she thought the lab approach was a good way for students to absorb the lesson—more so than a lecture. "I disagree," I said. "Good!" said Mr. NCTM in a disagreement-is-healthy tone of voice. I said that the same information could have been imparted directly while still challenging students to answer key questions all in a much shorter amount of time. Reaction: Silence. Mr. NCTM moved on to the next comment from another woman who in all seriousness and with no sarcasm intended said: "The teacher was very good at not answering the students' questions." There was unanimous agreement.

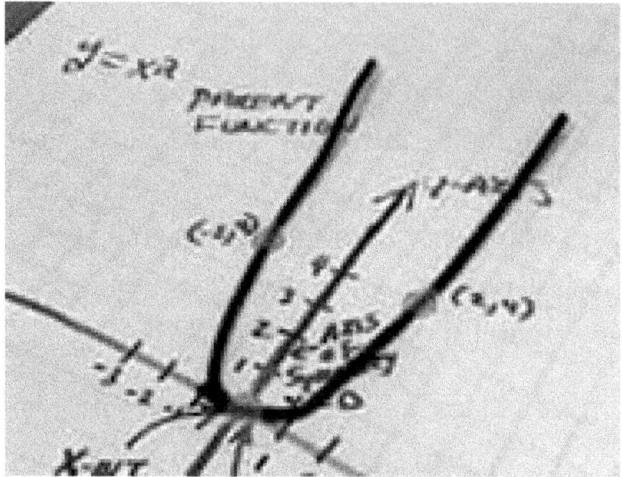

You can be very good at doing something that is absolutely horrible. And when you see teachers like these who are very good at what they do, if you didn't know any better you would try to emulate having students spend an entire class period bending pipe cleaners into parabolas and gluing them on poster boards.

6. More on Constructivism

Apparently my last missive ruffled some feathers, which I knew would happen sooner or later. It is one thing to express self-righteous indignation about Ed. school, but when it crosses the line into criticism of constructivist or "discovery learning", it's like a Congressman talking about revamping Social Security.

The terms "constructivist" and "discovery learning" mean different things to different people. To the Ed. school gurus as well as book publisher/snake oil salesmen peddling their wares to school boards who eat this stuff up (and make the final decisions on what textbooks to adopt) it means students construct their own knowledge out of whole cloth. To the more traditional-minded, it means the connection that students make between information given to them directly and applied in new situations, or which lead to new insights.

Students may remember having made a connection all on their own, but may not remember the guidance and information that a teacher or book imparted that got them there. There may be an "illusion" of pure discovery at work here: people see what they want to see. One interesting case in point is the TIMSS Videotape classroom study of math and science classes in other countries. When the video was released, constructivists said "See? See? Japanese students work in groups, are given challenging problems without instruction on how to solve them, and the student has to invent his or her own solutions."

But an interesting paper by Alan Siegel of NYU in fact shows just the opposite. (You can find his paper here: http://www.cs.nyu.edu/faculty/siegel/siegel.pdf.) Siegel describes the presentation of a geometry problem in a Japanese classroom and notes that the teacher provides a key theorem to students *prior to* presenting them with a problem to solve using that theorem.

The problem was quite good and since all of us in Mr. NCTM's class each have to present a problem to the class during the semester, I chose that one. I thought it would be interesting to see just how easy or hard it would be for the students in class to solve the problem given the theorem prior to the problem, just like in the Japanese classroom. In the video, the eighth graders were not able to solve it, even with that knowledge; they eventually got it through expert coaching from the teacher. Many constructivists do not seem to remember the teacher providing the theorem beforehand, nor that the teacher was a "sage on the stage" disguised as a "guide on the side".

So I presented the problem to the class, saying I would like their feedback on whether such problem is appropriate for eighth graders. After my initial presentation of the problem I told them I would give them three minutes to work on it, but not to feel they had to solve it—I just wanted to reconvene at that time and then discuss it as a class. (This is in fact what they did in the Japanese classroom). All fell silent and worked at their desks. (Note to adherents of people-working-in-small-groups: In our class, when we are given a problem to solve, most of us like to solve it in isolation. When instructed to work in groups, one person in the group generally dominates. My mind becomes paralyzed and I crave being left to my own devices.)

After about a minute, I saw that people were perplexed, not getting anywhere, and I suddenly realized that in my excitement: *I forgot to present the theorem they would need to solve the problem.* I apologized and called for their attention and explained the key theorem they would need.

Now, I fully expected that no one would solve the problem in the three minutes and I would have to be "guide on the side" and coach them to see how to apply the theorem, thus proving to all who believe in constructivism that students can still "discover" when given information directly. I forgot that my classmates all have a math or science background and are not eighth graders. Three of my classmates solved it within a minute and others were on their way. Nevertheless, my oversight in not presenting the theorem did reveal something important: As smart and experienced as my classmates are, no one was having any great insights into a solution until I presented the theorem.

I led a discussion about the appropriateness of the problem for eighth graders. The people who solved the problem immediately thought that perhaps I should *not* give the theorem and let them "discover" it. Others who had a tougher time with the problem said, well, if you did that, maybe you should coach them to come up with the theorem rather than expecting them to do it on their own. Or maybe giving them the theorem wasn't such a bad thing.

I suspect that the ones who had the easiest time were under the illusion that the theorem was superfluous and easily discovered. They forgot that a few minutes prior they were struggling until I told them what they needed to know. Just like people who in their memory believe they discovered all that was important in math. In short, anyone who was a constructivist at the beginning of the evening, was still a constructivist at the end of it.

The teacher in a Japanese classroom provides a key theorem to students prior to presenting them with a problem to solve using that theorem.

From the TIMSS Videotape Classroom Study, presented in "Understanding and misunderstanding the Third International Mathematics and Science Study: what is at stake and why K-12 education studies matter" by Alan Siegel, Courant Institute, New York University

7. A Good Swift Kick

One summer during college I had a brief stint working the night shift at an all-night drugstore in a rather scary section of town. On my first night, while waiting for someone to buy something or shoot me, the dead drunk security guard for the store came over and introduced himself. He put his arm on my shoulder and muttered something that I couldn't make out. I kept asking him to repeat himself which made him angry until he shouted: "I said you don't have to worry about anything with me around." I did not find this clarification reassuring.

This situation reminds me a lot of what I'm going through in Ed. school. I am confronted with explanations I can't quite comprehend, but whose clarifications upset me further. A case in point is the textbook we are reading in my math teaching methods class. The textbook is "Teaching Mathematics in Secondary and Middle School" by James S. Cangelosi. Excerpt from Chapter 4:

"Because mathematics is widely misunderstood to be a linear sequence of skills to be mastered one at a time in a fixed order, some people think teaching mathematics is a matter of following a prescribed curriculum guide or mathematics textbook. ..."

That would be me. Sorry, but I find that teaching the distance formula before delving into what is the Pythagorean Theorem, omits necessary logic and structure. Or teaching the quadratic formula first with derivation later, or in some cases, no derivation at all. That this type of "anything goes" technique with no regard to mastery is embraced by those who decry the practice of giving students formulas to memorize without understanding underlying concepts is also disturbing. Given how Cangelosi believes mathematics is "widely misunderstood" however, I would guess that I'm not alone in my beliefs. He goes on:

"Textbooks present information and exercises on mathematical topics, but typical textbook presentations are pedagogically unsound from a constructivist perspective. … Thus, textbooks should be used only as references and sources of exercises–not religiously followed page by page."

I think I've talked about constructivism enough for you to know my reaction to that. He concludes his rant with the following:

"Word problems from mathematics textbooks provide convenient exercises for students to experience some–but not all–aspects of real-life problem solving. With a real-life problem, students are confronted with puzzling questions they want to answer. Textbook word problems…present puzzling questions, but rarely are they questions students feel a need to answer."

This brings up the issue of just what a "real life" problem is and why it's different than the traditional ones the author eschews. Interesting that he feels students rarely feel a "need" to answer textbook word problems. I've been observing classes at a school in which the math teachers teach religiously from Dolciani's algebra textbooks (written in the late 60's and then revised in the 80's and were—and remain—very effective at teaching algebra to mastery). The students I observed at the school find the word problems in Dolciani challenging. In the spirit of full disclosure, these are honors and "gifted and talented" students, some taking algebra in the 7th grade. I tell you this for those of you who believe that mastery and higher order thinking skills come naturally to bright kids anyway and they feel a "need to answer" everything.

In addition to the pronouncements made in the textbook, Mr. NCTM handed out a one-page excerpt from a paper at the end of class a few weeks ago and said we would discuss it next session. It was an essay against the "traditional" word problems in algebra in which the unidentified author stated that such problems "convince students that there are no real applications of algebra, since the problems are so ridiculous." He gave an example of a work problem: John shovels snow from a walk in 4 minutes; Mary can do the same walk 3 minutes. How long will it take them to finish the job together? The author rails that no one can shovel snow that fast. Change it to 30 and 40 minutes if it bothers you so much; the concept is still the same. But the author is not concerned with that. The author of the essay finds algebra problems to be such that students will ask, "Who cares what the answer is?"

Like hearing what the drunken security guard at the drugstore was trying to tell me, I dreaded what Mr. NCTM would say about the essay. I fully expected a facilitated class discussion ending with the conclusion that short-term relevance trumps content and mastery—problems that are messy and time-consuming like finding the best long distance telephone plan are much more instructional. To my surprise, the essay was not discussed. Mr. NCTM said only that it was written by a "very smart mathematician" at University of Chicago in the 1980's, a man by the name of Zal Usiskin. For those who don't recognize the name, Zal Usiskin was a major player in the development of the Everyday Mathematics program which is used in K-6.

He is most likely responsible for the following which appears in the Teacher's Reference Manual for that program: "The authors of Everyday Math do not believe it is worth the time and effort to develop highly efficient paper-and-pencil algorithms for all possible whole number, fractions and decimal division problems.... It is simply counterproductive to invest hours of precious class time on such algorithms. The math payoff is not worth the cost, particularly because quotients can be found quickly and accurately with a calculator."

If I may add my own clarification to both Cangelosi (who has a masters in math) and Usiskin (who has a PhD in math education): In other words, the U.S. doesn't need to produce scientists and engineers when we can hire them more cheaply from India and China where traditional word problems are presented—and solved with alacrity. Whether Messrs Cangelosi and Usiskin need good swift kicks is something I will let the reader discover in true constructivist spirit.

8. Glasnost, Perestroika and Graphing Calculators

After last week's missive quoting from Dr. Cangelosi's textbook, I expected he would have left a comment expressing his eternal gratefulness for the exposure I gave his book. But he lives in Utah, where the state legislature there recently adopted a resolution that calls for the Utah State Board of Education to give Utah's math standards an overhaul. That may have him worried and I'm sure that's why he hasn't written.

From what I see and hear in Ed. school, Dr. Cangelosi doesn't have a thing to worry about. The milieu-controlled Ed. school environment of discovery/inquiry based/NCTM-standards-based/constructivist-based/brain-based/knowledge-based/critical thinking-based/ and higher order thinking skills-based learning is ever expanding.

Indicative of this brave new world is a comment that Mr. NCTM left on a lesson plan I turned in—an assignment that called for a lesson which made use of technology.

My lesson plan had students explore the graphs of quadratic equations using graphing calculators. I borrowed heavily from exercises in a math book by Gelfand on functions and graphs. In one of Gelfand's non-calculator based exercises, he asks the students to graph $y = x^2$, and then $y = 5x^2$ and asks "What scale unit would have to be taken along the axes in order that the curve for $y = x^2$ could serve as a graph of the function $y = 5x^2$?" Students are to provide a rule linking the shape of curve to the x coefficient, based on their answer to the scale unit question. Mr. NCTM wrote in the margin: "This is just the kind of 'discovery' learning that you have rebelled against."

His comment reminded me of the movies made in the Cold War era in which a staunch Soviet leader says to the American hero: "Perhaps, comrade, we are not so far apart as we thought." I believe what he saw in my lesson plan was indeed a form of discovery learning which involves "scaffolding" as a means for students to make cognitive leaps and "discover". Scaffolding refers to incrementally increasing the difficulty of problems by introducing subtly varying the information provided in a problem. In the problem above, a student would need to think about what the question is asking and work with scale a bit, and in the process discover how a coefficient in the standard equation of a parabola could widen or narrow the shape. As such, the technique allows (maybe forces) students to apply prior knowledge to a new situation or problem. So perhaps in this sense he is correct that we are not so far apart. But in other areas, despite the optimistic nature of his remark, I do not feel we are really as close to glasnost as he would like me to believe. Perhaps he sees my acceptance of scaffolding as a chink in the armor on the way to get me to see things the NCTM way. I do not, however, accept the broad brush, ill-posed, open-ended problems that force students to make a hasty discovery in a "just in time" manner and which is forgotten after the bell rings.

His view of me as a dissident in need of enlightenment comes from things I say in class, most recently in a class discussion on the role of graphing calculators in math education. The discussion started in the usual manner: get in small groups. In my group was the fellow with whom I had an argument about state standards the first class. I've grown to like him; he's very young and full of opinions and enjoys being contrarian. For many people in their twenties being contrarian is a quest for identity until marriage, work and humility take over, and not necessarily in that order. In any event, for this young man none of these things have yet kicked in.

Mr. NCTM facilitated the classroom discussion. While we are not Luddites in our class, and can appreciate the value of graphing calculators in teaching, we also saw problems. After several minutes the whiteboard was filled with issues including overdependence, obscuration of concept, and interference with conceptual mastery. After some discussion of the pros and cons of graphing calculators, Mr. NCTM decided to change tack on us and asked: "Do you think they are introduced too early?" (They are introduced as early as kindergarten in some programs.)

Our answers were going in the direction of "yes", until my young contrarian friend spoke up and said, to Mr. NCTM's obvious delight, that he really couldn't see what was the cognitive value of teaching students the procedure for multiplying 36 x 7 when calculators were available. I was unable to keep my mouth shut. "Don't you think that students need an understanding of basic procedures and that place value is an important concept?" "Why?" he remarked and went on to the uselessness of learning long division at which I drew the line and said "How can you say that? Don't you think the distributive property is worth talking about?"

"Who cares?" he offered constructively.

Mr. NCTM was enjoying this debate immensely. Dialogues such as these apparently feed into his fantasy that he's actually teaching in a real grad school in a real program.

Mr. NCTM took over and allowed that there was *some* value in teaching the long division algorithm and perhaps *some* value in multiplication algorithms, but after that, it is just so much tedium. "There are some who feel there should be no pencil and paper calculators in classrooms at all; you either do it in your head using estimation or you use the calculator. It breaks my heart when I see kids writing down

32 divided by 2 and solving it as a long division problem." It breaks my heart too. Students used to be required to practice problems such as these until they could do it in their head as he would like to see. Such problems used to be called "short division". Apparently, Mr. NCTM sees long division as causing this problem, not the calculator.

"Let's put it this way," Mr. NCTM said. "If I saw a student who was not able to perform the division problem of 168,514 divided by 384, I would not view that as a reason to hold him back from taking algebra." Well, if it were me, I would first want to know why he couldn't do the division problem and then what else he couldn't do.

Which tells me that Mr. NCTM and I are a long way from perestroika.

9. The View from Behind the Counter

Greetings and thanks to my many fans and well-wishers for their undying support, encouragement, wisdom and guidance. I am happy to say that my Math Teaching Methods course is at long last over. For those of you wondering how I've done, I'm getting an A in the course. I have not kept secret from the teacher my opinions of how math should be taught and though we disagree, he has offered me this final email message: "I have very much enjoyed sharing the classroom with you.

Your insights and comments have been extremely valuable, and your willingness to communicate your point of view has served as model behavior for your classmates. Thank you very much."

There are some positive aspects to Mr. NCTM I'd like to mention. He has had 30 years of experience teaching high school math, knows quite a bit of math, has a good sense of humor, and has provided my class excellent advice regarding classroom management issues, and other things such as how much material to cover in one lesson plan, and what concepts students find difficult. Our difference in opinions has not influenced the grading of any of my work. (Note: He does not yet know about this column, so if you wish to tell him about it, please wait until after the grade is in the transcript.)

My classmates are quite bright, and if I led you to believe they are all dyed-in-the-wool constructivists, let me set the record straight. Only one is gung-ho, three or four are willing to give it a go, and the rest keep silent. The young man who is contrarian and with whom I got into arguments is the son of a mathematician, and is quite demanding of rigor. I suspect that as he gets older his love of rigor will point him in a direction quite opposite that of constructivism.

So where does all this leave me and Mr. NCTM? Right where you found us. My idea of "discovery" is quite different than his. Take the problem of finding a formula for the sum of interior angles of a convex polygon which we discussed in class. The solution hinges on the fact that the sum of the angles of a triangle equals 180 degrees. You could approach teaching this lesson by guided discovery, and show how to split up a quadrilateral and pentagon into triangles to derive the sum of the interior angles. After several minutes of discussion, some students may identify key patterns and the teacher could wrap it up. I suggested this to Mr. NCTM, remarking that it really wasn't giving away the store and there was still some discovery involved.

He nodded acknowledgement but continued to "guide" our class to the constructivist approach: 45 minutes of discovery including having students actually measure the angles of various convex polygons with a protractor, and after accounting for error in measurement, making conjectures and seeing "patterns". One fellow student asked why one would do that when in fact geometry was about deductive reasoning and learning to reach conclusions about measurement without the aid of actual measurement devices. Mr. NCTM said students should be given free rein to discover the superiority of the deductive method.

For fear of being forever branded as a blog poseur, I hesitate to say whether my approach would be called "direct instruction" or "guided discovery" or Vygotsky's Zone of Proximal Development and its related term, "scaffolding." I just offer that the type of discovery my approach entails is what I and others in my age group had growing up. Given a choice between giving students 45 minutes to reach an "aha" experience, or 5 to 10 minutes, I and others like me opt for the latter.

It is interesting how one imbued with the NCTM philosophy of teaching views the world. In a discussion about textbooks, we asked him what books were his favorites. To my surprise, he liked Foerster's algebra, and Jacobs' geometry texts—two classic and very traditional texts. Nevertheless, as good as they are, in the NCTM view (and Dr. Cangelosi's as well), textbooks are mere resources by which to provide exercises and problems for the "lower level" algorithmic skills. The real teaching and learning is in constructivist lessons. In answer to why there are bad textbooks out there, Mr. NCTM remarked that the NCTM standards haven't made their way into the textbooks as much as he would have hoped. That the NCTM standards themselves may be a reason for bad textbooks is probably not something he has even remotely considered—just a guess.

This brings me back to the drugstore in a scary part of town where I once worked. Something else happened there that is relevant to the differences between us. While working the day shift, I was given instruction by a seasoned veteran on how to operate the cash register and not take any guff from customers. The customers were the enemy in her eyes—not to be trusted, not to be friends with. We ran the show.

Because I was also given night shift duties on occasion, my father was not thrilled with me working there, and I was forced to find another job. About a week after I left the drugstore job, I returned as a customer. I saw my old trainer/mentor behind the counter and I waved to her. Her eyes passed over me like they did everyone else on our side of the counter. I was now the enemy, unrecognizable as someone she once counselled, and on the other side of a gap that would never be bridged. I am grateful to both teachers and wish them well. But I know on which side of the counter I will remain.

Letters from Huck Finn

-Kemble

Part I: Starting Out

1. Who I Am, Where I've Been and Where I Think I'm Heading

For those few of you who know me, you probably know me by John Dewey, a nom de plume that I used to write a series of letters about education school on a blog called Edspresso mostly in 2007. I am a working stiff who will be retiring this year after which I will pursue an après retirement career of teaching math in middle or high school. I have finished all my courses and only have to complete my student teaching in order to get my certificate.

I decided to resurrect my letters but the name John Dewey seemed to belong to a different era and didn't feel quite right. I'm now in an in-between mode, reflecting on all that has come before, and what will come after, floating down a river on a raft, having a vague idea of where I'll end up. I've decided therefore to call myself Huck Finn. I welcome any of you to either listen to my tales of woe or tell me of your own. For those of you who choose to climb aboard the raft to talk of your enslavement by the education establishment, I will call you by a name designed to not in any way invoke the ongoing controversy surrounding the famous novel by Mark Twain, nor to make my kind hostess Miss Katharine nervous. I will call you James.

As followers of the John Dewey letters know, this river goes through territory that is decidedly divided between conflicting theories of education. The divisiveness has not subsided any since I stopped writing the first round of letters and it doesn't look like it will end any time soon. School districts, school boards and the administrations in their grasp remain smitten by the lure of the promise that math (or any subject for that matter) need not be taught in any kind of logical sequence, and that whatever mastery of facts or procedures is needed can be learned on an as-needed basis because process is more important than content.

Some say that the divisiveness in education has been around for quite a while but I have to say that free form pondering and group discussions were the exception and not the rule back in the 50's and 60's when I was in school. When open-ended discussions did occur they were short lived—like the time in my 8th grade science class when the teacher, Mrs. Cohen got off on a tangent of what came before the universe was created. I don't recall how we got off on that particular tack. I think she sensed that a full exploration of the topic might take more time than she had, so she brought the discussion to a memorable close by announcing that such things were beyond the capability of the human mind and that there had so far been only one human being capable of understanding these origins. We all thought she was referring to Albert Einstein and were therefore surprised when she announced that this person was (wait for it): Rod Serling.

Being that it was 1963 and Twilight Zone was at the height of its success, we had no problem accepting that. With that business done, Mrs. Cohen got back on track and we didn't have to write any essays comparing Einstein to Serling, or work in groups to construct shoebox dioramas of the creation of the universe.

But now we are in an age of education by collaboration, by small groups, by a student-led and teacher-facilitated inquiry-based approach. My educational psychology professor at Ed. school was the personification of this practice. She bore an uncanny resemblance to Meg Ryan and had taught middle school science for 15 years. Her classes were chock full of group activities, designed to help us learn the material as well as to become conversant with the various group activity techniques.

I never learned her stance on Rod Serling versus Albert Einstein (or even Stephen Hawking) but she was big on student engagement and "hands on" learning. The extent to which she believed that "hands on" or inquiry-based or student-centered learning should be grounded in specific content that is explicitly taught remained a mystery. She left clues throughout the semester, however, that she leaned toward the Ed. school thoughtworld that students should construct their own knowledge.

She definitely believed in student engagement as a key to motivation. I found this out when she told us about a particular "back to school" night for parents during which she demonstrated how she taught her students about the difference between force and pressure:

She lay on a bed of nails with a rock balanced on her stomach and had a fellow teacher smash it with a sledge hammer. All of us were suitably impressed. Yes, she built the bed of nails herself. And no, she didn't demonstrate it during our class.

I don't expect that I will use the bed of nails technique with my future students. Nor do I expect to do half of the engaging engagement activities that I learned about. For those who are curious, I got an A+ in the class. I think it was because I used the term "Piagetian disequilibrium" in one of the papers I had to write. I also learned that if you have to make it look like you believe in students constructing their own knowledge, you can talk about Vygotsky, and the Zone of Proximal Development, and scaffolding, and people will see what they want to see. I thank Meg Ryan for teaching me the art of camouflage.

Well, it's getting late, and Miss Katharine is looking like she's nervous so I'll wrap this up. My application to student teach is due in a week and I have to write a goals statement that sounds like I won't tell my students what they need to know. It has to sound like I mean it. This might take me a while.

2. Drifting Down the Great Divide, and the Sunrise in the South

It has been a year since my last confession, otherwise known as a letter. In the meantime, I completed my student teaching in December. I'm the same person who wrote a series of letters under the name of John Dewey some years ago, chronicling some of my experiences in Ed. school.

After my introductory missive as Huck Finn a year ago, some readers were downright angry so I'll just say that with respect to the professor who bore a resemblance to Meg Ryan, I got a lot out of her class, and have nothing against someone lying on a bed of nails while someone smashes a rock on her stomach with a sledge hammer on back to school night or any other night. It's a lot better than most of the back to school nights I've attended. My daughter's 7th grade speech/drama teacher's greeting was that she just got back from Bahrain and would rather still be living there than teaching here. I don't know what was more remarkable: the fact that she said that, or the fact that the parents in the audience nodded their heads in appreciation.

I have chosen the name Huck Finn since it seemed to fit better than John Dewey for someone drifting along the pedagogical, political and cultural river that crosses the battles over how best to teach math. The battles over math teaching remind me of something that happened years ago. I am a creature of habit, and throughout high school I walked to school every day, which meant walking eastward every morning. When I started at college, I consequently assumed without thinking about it that my walk to my first class was eastward. I was surprised therefore one day to see the sun rising in the south.

For a moment I considered what could have caused this, before I realized that I was walking northward. My experiences in education are similar. I meet people who see a sun rising in the south, can explain it, and will not entertain any suggestion of where true north is.

I will reveal that I live in California, but had completed my Ed. school course work at a school back east that allowed me to do my student teaching in California. My effort was coordinated with a local university here who said they would place me in a school to do my student teaching. By mid-

August, I still hadn't heard about where my placement as a student teacher would be.

I told them that the school back east requires me to put in 15 weeks of student teaching and the schools where they wanted to place me would start next week. There was a flurry of activity suddenly. I was told to write a bio which they would circulate. My bio mentioned that I had majored in math and had been in the Michigan Marching Band—not that these two things are related but I wanted to show that I was a team player.

The next week, I was told to report for an interview at a junior high school about an hour's drive from my house. The school's website listed the textbooks they used. I was disconcerted to see that for algebra, they were using College Preparatory Math (CPM): Algebra Connections. It is a discovery-based program and one that caused a significant uproar when it was used in Palo Alto in the early 90's.

I met with Tina who would be my supervising teacher and had taught for 10 years. She mentioned she had a son in high school who was in the marching band and she herself played in the band at the community college. I thought perhaps my mention of the Michigan Marching Band had caused her to select me. "I guess you know from my bio that I was in the Michigan Marching Band," I said.

"What bio?" she asked. The world of education is seldom as organized as one might believe.

In the course of our conversation she revealed that she liked the standards of the National Council of Teachers of Mathematics (NCTM and was also happy that California had just adopted the Common Core standards for math. "Do you know those?" she asked. My feeling about Common Core is very similar to that of NCTM's but I left it at: "I am familiar with both."

"So they placed you in a middle school," my teacher said as if confirming that I had a disease that had no cure and was both fatal and painful. "Are you OK with that?"

"Actually, I requested middle school," I said.

"Why?"

"I'm out of my mind," I replied. She thought this was pretty funny and I was tempted to leave things at that, but I added that middle school was often the last chance students had to get proficient with fractions, decimals, percents and other concepts that they may be weak on before the onslaught of algebra.

"Exactly!" Tina said. "I taught high school for a year and was so heartbroken when I saw seniors who were still taking algebra 1 and not passing because they didn't know their basic math. I decided that middle school was where I should be."

This was beginning to show some promise. She said I would eventually be teaching three periods. Two periods were pre-algebra which used a standard, traditional-style textbook. The other period was algebra which used the text known as College Preparatory Math (CPM): Algebra Connections. "It's discovery-based," she said with some excitement and perhaps an expectation that I would rejoice in this. I looked around the room at the desks pushed together in groups of four, and tried to believe that somehow we would all know and agree where true north is, and that the sun rises in the east.

3. Navigating Between Algebra and Tomato Paste

I student taught for 15 weeks at a school I'll call Aragones Junior High. The school is located in an agricultural region known primarily for its strawberry fields. It is in a residential neighborhood surrounded by hilly farmland. Some of the farms have cows on them, but most grow strawberries. The make-up of the school is about 96% Hispanic students, many of whose parents work in the fields I would be passing on a daily basis. During my time there, the students sometimes appeared to me as adults, sometimes as kids, and other times as people caught in between

The first weeks of my program I would be observing the three classes I would be teaching; by the fourth week I would begin to teach some classes. On my first day Tina, my supervising teacher, asked me to circulate among the students during the algebra class and answer questions or offer help as needed. The algebra class was a discovery-based class that used the CPM textbook (College Preparatory Math). The problems the students were working on in their groups of four were "guess and check" problems. The book spends more than half of the semester on these types of problems and even provides instruction in how to set up tables of values to maximize the efficiency of an inefficient process. Tina even gave instruction and pointers on how to do this during the class. She's a good teacher and explains things well, which bolsters a point I've made in my previous incarnation as John Dewey: It's possible to do something horrendous tremendously well.

A moment of full disclosure: I went to school at a time when math was taught in a traditional manner in K-12. Despite my average intelligence and despite claims from various quarters that such method was more destructive than typhoid fever except for very bright people, I managed to learn enough to allow me to major in the subject. The algebra book I used started almost immediately with how to express words as algebraic expressions, and to use that skill to set up and solve equations. The book briefly discussed the "guess and check" technique which at that time was called "trial and error". It illustrated how a problem could be solved using trial and error, and then how the same problem could be done quickly and efficiently with algebra which then remained the focus of my algebra course.

My class, however, was stuck with "Guess and Check". A boy named Rudy asked me for some help with a problem. Rudy was fairly bright and his group mates seemed to rely on him to get through the problems. The problem was as follows: "In making a batch of soup, the number of cans of tomato paste was five more than twice the number of cans of noodles. A total of 44 cans were emptied into the soup. How many cans of each ingredient did the team use?"

I was mindful of the not-very-optimistic warning I got from the local university which placed me at the school: "You are a visitor/guest in the classroom. If there are any differences of opinion with the teacher, things have to be done as she wants them since it is her classroom." Thus, I strived to adhere to the guess and check nature of the assignment. I knew from watching a pre-algebra class earlier that students learned how to translate English expressions into algebraic ones. I figured that these students knew how to do that at the very least. But while CPM touts itself as connecting knowledge, connecting to what they learned in pre-algebra was apparently not on the agenda.

"OK," I said. "If you have twice the number of cans of noodles, how can we write that?"

They stared blankly at me.

"How do we write 'twice'?" I asked. "What number do we use?"

Rudy brightened and said "Two".

"Yes, two; so if I have four cans of noodles what's twice that amount?" Rudy thought a bit while the others in the group looked at him.

"Eight."

"Right! So if I call a can of noodles 'x', what's twice that amount?"

Rudy thought again. "2x?"

Now we were getting somewhere. "OK, so if the number of cans of tomato paste is 5 more than twice the number of cans of noodles, how do we say that?"

Rudy thought for a moment. I expected to hear "2x + 5" but instead, he said: "Paste."

I asked the others. They all said "Paste." The others now started to giggle. What I didn't know was that the students had been setting up guess and check problems with headings like "cans of noodles", and "tomato paste = twice the number of cans of noodles + 5". Rudy was stuck on "paste" as an answer and the more he said it, the more I tried to get them to use letters rather than words. The futile conversation with Rudy was making them all laugh. They tried not to, but that only made it worse. They all spoke English well, but for all practical purposes we were speaking different languages. For lack of experience and anything better to say I told them,

"If you want to joke around, you're wasting your time and mine," and moved to the next group.

Tina told me later that the class wasn't ready yet to translate into expressions using "x". "You don't know what 'guess and check' is about yet," she said. The motif of old school teacher meets the modern method then became the basis for her retelling the story to others in the teacher's lounge. "Guess and check isn't easy to teach," she said consolingly. "The book eventually does connect guess and check with equations, and then the light bulbs come on when they put it all together."

I was about to say it might be a lot easier just to teach them algebra. But I thought it probably would be best to keep my mouth shut, so that's what I did.

4. Thinking about Inverting, Multiplying, Understanding, and Hanging

Miss Katharine gave me an earful after my last letter. She was bent out of shape over my statement that I am of "average intelligence". I tried to tell her that enough people believe I'm an idiot or of otherwise low intelligence because of my opinions on education to allow me to get away with saying that. She mumbled something I couldn't hear about who the real idiots are and then told me to "clarify matters". So here goes:

I may be above average in intelligence, but I am neither a math genius, nor am I brilliant. I make this distinction because I do not believe that traditional math serves only those people who are brilliant and/or are pre-destined to study mathematics. I believe it provided many people with sufficient preparation to take calculus in high school or college.

Some think that the old ways of teaching math were just rote memorization, with no understanding. To be blunt, I happen to think that's a pile of crap. To be more refined, I happen to believe that procedural fluency leads to understanding; once you're able to do certain procedures, it's easier to understand why they work. Obviously, there is a lot of back and forth on this. Tina, my supervising teacher, believes that students shouldn't just be told how to do procedures; they should understand the "why." She is, however, a firm believer in having students do many problems to gain procedural fluency.

"Do you think you'd like to try teaching a class?" Tina asked me one day. I said I was ready, and we decided on a day that I would do it, which turned out to be the day that fractional division came up in the pre-algebra class. "That one's a doozy," she said. "Are you sure you want to take that on?"

"It's my favorite topic," I said.

She was intrigued with this and told me about people not being able to explain how the "invert and multiply" rule works. She is correct. In discussions, arguments, and brawls over "understanding", the invert and multiply rule for dividing fractions is the poster child. For those who view understanding as paramount, the fact that many people are unable to explain why the rule works is considered as another piece of evidence that traditional methods have failed.

"The students already learned how to do this, in fifth grade," she said. "But for pre-algebra I really like to make sure they understand why it works." And I agree. Explaining the derivation of the invert and multiply rule is an appropriate topic for a pre-algebra class.

I suggested that since we had just finished the chapter on solving one step equations, we could build on that knowledge. Since 7/8 divided by 3/32 equals some number x, then we know that $(3/32) \cdot x = 7/8$. And since they know that 3/32 times 32/3 equals 1, we can then isolate x by multiplying each side by 32/3, and voila: $x = \dfrac{7}{8} \cdot \dfrac{3}{32}$

She thought a moment and said she likes to use the "complex fraction" method. So we write the problem $\dfrac{7}{8}$ divided by $\dfrac{3}{32}$, as a complex fraction:

$$\frac{\dfrac{7}{8}}{\dfrac{3}{32}}$$

To simplify, we want to get rid of the $\dfrac{3}{32}$ in the denominator. The easiest way is to multiply by the reciprocal, $\dfrac{32}{3}$ which makes the denominator equal to 1. But if we multiply the denominator by $\dfrac{32}{3}$ we have to do the same for the numerator. This sequence of steps ends up looking like:

$$\frac{\dfrac{7}{8} \times \dfrac{32}{3}}{\dfrac{3}{32} \times \dfrac{32}{3}} = \frac{\dfrac{7}{8} \times \dfrac{32}{3}}{1} = \frac{7}{8} \times \frac{32}{3}$$

And there it is; invert and multiply in all its glory. To ensure they really see it, she has students use this method in their homework problems rather than simply inverting and multiplying. This appealed to me because of my belief that procedure leads to understanding, and that the repetition of doing it "the long way" at the very least would make them grateful for just inverting and multiplying. Every time they did it the short way, they would remember with thankfulness not having to go through the rigmarole—which just happened to be the derivation.

Tina taught the lesson to the first period class, and then had me teach the same lesson for the third period class. Unlike the discovery-based algebra class, the pre-algebra classes were taught in a traditional manner. Nevertheless, when my time came to teach, I was a bit nervous, a bit rushed, and probably stood in front of the whiteboard too many times, blocking my own writing.

To make sure they connected with what they already knew, I started off with a question: "How many of you remember how to divide a fraction by a fraction?" Many hands went up. I called on Emilio. I waited. He said nothing.

"Well, Emilio, how do you do it?"

"Oh, I thought you were just asking if we remembered how to do it so I raised my hand." He will either go into math or law, I remember thinking.

"Well, as long as we're here, why don't you tell us?"

"You flip the second fraction upside-down and you multiply," he said.

"Correct," I said. "But now we're going to find out why it works." I went through the explanation, asking questions as I did so to make sure they were following. I gave them some fraction division problems to do, instructing them to do it as I had done on the board.

In the end, the students knew what problems are solved by fractional division as well as the procedure. But I'd also say, that in the days that followed, once they were again allowed to use the short way (invert and multiply), the derivation did not matter much and probably only a few could reproduce it if asked. That doesn't bother me. I'd rather they know how to solve problems than be able to reproduce an explanation they don't fully understand for a procedure they cannot perform. A teacher friend of mine told me not to say that too loud or they'll hang me. Well then, I guess when judgment day comes in the education world you'll find old Huck hanging from a tree.

5. I Learn the Ropes and Give Things Some Serious Thought

During my student teaching, I tended to stay away from discussions of educational issues. Judging by the reaction to my last letter, discussion of educational issues can result in disagreements expressed at great volume. I stayed away from politics and religion as well. I did venture a bit into football, though, which is not without its risks. The principal was a Notre Dame and Michigan State fan, and I went to University of Michigan. She didn't like Ohio State, though, so we got along just fine.

While I don't consider student teaching similar to football, I was often reminded of my experience in the Michigan Marching Band. I was in the Band when it was directed by the legendary perfectionist William D. Revelli.

The Band's high step marching was physically demanding. I had enough wind to enable me to march but I couldn't play a note on my clarinet all season. I kept that my dirty little secret, but when I rejoined the band the next year, I was delighted to find that I had the physical endurance to march and play at the same time, and incorporate all the nuances and instruction hollered at us during rehearsals.

I hoped that what happened in the band would happen in teaching: that I would be able to simultaneously teach and automatically do with ease what I now had to be told to do. My supervising teacher, Tina, often interrupted my teaching. She would whisper things like: "Circulate; see what they're writing—you're chained to the desk." "You take too long to answer the question; go for the big idea." And in the discovery-based algebra class: "You're telling them too much. You have to ask them questions; if you do it all for them, they'll never learn." This last one was perplexing: was it an issue of proper questioning—an important skill to learn—or was it an application of the Ed. school adage that if students aren't struggling, they aren't learning?

I began teaching first and third period pre-algebra classes my fourth week. Third period was an honors class. Though I liked teaching both, I found the honors class more motivated and responsive than the first period class. I admit to feeling guilty in saying that the honors class was more fun to teach, though I suspect I'm not alone. Some teachers, given a choice, might not want to teach struggling students, but then fall prey to being called elitist. Neither teachers nor schools want to be accused of that. So now in the name of equity for all, schools have mixed ability/performance classes. I'm happy to say my school grouped students by ability/performance. I don't believe honors and gifted classes are elitist. I think all students need attention.

That said, first period was a bear to teach. I was always nervous before it began; fifteen minutes before the bell I would have to run for the bathroom. Once the bell rang, and I swung the door open, I would relax. As students entered the classroom, the following exchange would continually take place:

Student: "Will we need our books today?"

Me: "Yes, get your books and warm-up folders, please."

After answering this same question about five times, I would eventually tell them to ask their neighbor, thus attempting to make good use of the groups of four desks clustered together.

Because of the non-responsiveness of the first period class, Tina and I decided to have them take out a sheet of paper first thing, to write down specific items (vocabulary, formulas), and then have them do a problem. "With this class you've got to get them doing something," Tina told me. "Otherwise they'll fall asleep on you. And you've got an evaluation coming up."

She was referring to the evaluation by the local university person who ultimately would give me my grade. He was a kind man who had taught high school math for over 30 years. I did well the first time, but that was the honors class. He now wanted to see me with the first period class.

The day he came I taught a lesson on inequalities. I had the class write down the symbols for less than, less than or equal to, greater than, greater than or equal to, which they already knew. I then added the terms: "Not as much as", "No more than/at most", "More than", "No less than/at least". I gave examples and then showed them a picture of a sign that you see at amusement parks that read "To go on this ride you must be at least 48 inches tall." I asked "If H is the height of a child, how would we write in symbols that the height, H, meets the requirements?"

Dead silence.

"For example," I said, as if nothing was wrong, "Would we write it like this?" I wrote H < 48. The class said "No-o-o-o."

A boy in front named Francisco said softly "H is greater than or equal to 48."

"Could you say that louder?" I asked. Francisco shook his head. Someone else gave the right answer.

"Yes," I said and turned to Francisco. "You see, you were right! You have to have the courage of your convictions. In math, people often disagree. And sometimes they express these disagreements at great volume." I stopped my pep talk there, thankfully. I didn't need anyone falling asleep during my evaluation.

I continued the lesson, students stayed awake, and I received a good evaluation which made Tina happy. Later during 4th period which was her "prep" she talked about things to come. "I think pretty soon we'll have you start teaching some algebra classes. Eventually, you'll be in charge of all three classes—doing the planning, quizzes, and tests. And the discipline—it can't just be me."

I excused myself and ran for the bathroom. The idea of being fully in charge was overwhelming. If I found two classes to be demanding how would I handle three? Relax, I told myself. I would never get that far. Tina would be watching every move. Sooner or later I'd get into a discussion with her about math education and we express our opinions at great volume. I'd be politely booted out and advised to do something normal for my age, like be a Walmart greeter.

6. My Teacher Leaves, and I Become a Gunslinger

Things were looking pretty rosy by the third week in October. I was at the halfway point in my student teaching and had made it through two evaluations and was getting the hang of things. But then my trip down the river took an unexpected turn. I was getting the first period class through the morning routines on a Monday when Tina's cell phone rang and she went outside to answer it. I could hear her voice through the door; it sounded like she was crying. She came in and told me she had to leave. "It's an emergency. My father's in the hospital. Can you handle the class?"

"I've got it," I said.

"They'll try to get you a sub. I've got to go." She grabbed her purse and ran out the door. I found myself without my teacher, facing 30 or so students waiting for me to do something. I now had to explain zero and negative exponents and get my first period class to follow along with the explanation the text book had presented. This was clearly not going to be easy.

The way I learned was based on the rule for dividing powers: $\frac{a^m}{a^n} = a^{m-n}$. Suppose you have $\frac{a^3}{a^3}$ (with a not equal to 0). Then it equals 1, because any number divided by itself is 1. But using the rule for dividing powers, it also equals a^{3-3} which is a^0. So $a^0 = 1$. Similarly, for negative exponents, something like $\frac{a^2}{a^3}$ is easily shown to be $\frac{a \cdot a}{a \cdot a \cdot a}$). By cancelling a's, you're left with $\frac{1}{a}$ and by the rule for dividing powers, a^{2-3} is a^{-1}. So, it's reasonable to say that a^{-1} equals $\frac{1}{a}$ which equals 1/a. This all made perfect sense to me when I learned it years ago. I liked the idea that the application of one rule led to another. But the textbook I was using now took a different approach. The authors seemed to think that deductive reasoning didn't teach the right "habits of mind" given the prevailing belief that "math is about patterns". So they teach students to use inductive reasoning in math.

Their explanation amounted to using a place value chart (thousands, hundreds, tens, ones, tenths, hundredths, etc.). I first showed that the thousands column can be written as 10 x 10 x 10 or 10^3. "Moving to the right of the thousand's place, what's next?" I asked. A few students responded: "One hundred". I wrote 10 x 10 = 10^2, and pointed out that this was 1,000 divided by 10. "And the number next to 100?" I asked. "Ten!" a few more said. "And look at our exponents. We had 10^3, and 10^2, what's this next one going to be?" "Ten to the one" they said. I then asked what the number next to the ten's place would be. "One," they said. I pointed to the pattern of exponents, and said "Three, two, one…what's the next exponent?" Someone said "Zero." And there you have it: by the pattern, 10^0 is 1.

Continuing in that fashion, I showed how when we move further right we get into negative exponents: 10^{-1} is $\frac{1}{10^1}$, 10^{-2} is $\frac{1}{10^2}$ or $\frac{1}{100}$ and so on. From this, students were to take it on faith that this pattern applied to powers other than ten, like 5^{-2} is $1/5^2$. I was fairly certain the explanation wasn't going to stick with them for more than two minutes—if that. So I had them write down the rules and relied on what can either be called an axiomatic definition or what the authors probably felt they were avoiding: rote.

My next challenge came when I saw a paper airplane fly across the aisle. Judging by where it landed, I had it narrowed down to one of two boys: Eliseo or Cesar. The class was silent, watching to see what I would do. I knew I had to do something. Up until now I had never made anyone "do a card"—a punishment in which a student had to copy what was written on a colored card kept in a folder at the side of the room: a treatise about the value of education. Increasing amounts of card punishments could get students a referral to the office, a parent-teacher conference, or even a suspension.

I'm not sure what prompted me to do this, but I walked down the aisle between Eliseo and Cesar in the manner of a gunslinger—slow, sure, and looking for danger. The closer I came to Cesar, the more nervous and fidgety he became. I made my choice. "Cesar, do a card," I said. He fairly jumped out of his seat and took to the task; he even looked relieved. The class seemed impressed with my feat. They seemed much quieter after that.

The day moved on. By third period—my last one—I was still on my own. The class followed the exponent lesson, but Manuelo, one of the brightest students in that class, asked the question I had asked long ago when I first saw zero powers: "How can anything be raised to the zero power? And why would it equal one?"

The class looked at me like my first period class did when Cesar threw the paper airplane. "Well," I said, "sometimes in math we just say something is a certain way because it fits the pattern." "OK, but I still don't get it," Manuelo said. The axiomatic approach was not sitting well with him. Nevertheless, he worked with the concepts of zero and negative exponents with the rest of the class.

During the "prep" period (fourth period), the phone rang. It was Tina.

"How are you doing?" I asked.

"I'm OK. My father's had a heart attack. It's extremely serious."

"I'm very sorry," I said.

"So I may be out for at least another week. I talked to the principal; you'll get a sub tomorrow. What did you do today?"

I told her about the exponent lesson. "Oh, we never do it the way the book does it. We teach them the rules for multiplying and dividing powers; you know—$3^2/3^2$ equals 1 and also 3^0?"

"I'm glad we're having this little chat," I said.

"Sorry," she said. "There wasn't time."

"I know, I know. You don't have to worry. I'm doing fine."

"I have full faith in you," she said. "I know you can do it."

We both knew we were lying to each other, but sometimes that's all you've got to get you through the day and night.

7. Trying to Hold On and Another Guest on the Raft

I was thinking of not writing any more letters since it seems no one is reading them. Miss Katharine just about had a heart attack when I told her this. She said a lot of people are reading them and if I am going to quit writing them, why in heaven's name after telling my readers that my teacher left for a family emergency and I was on my own, would I leave them hanging? I had to agree with her. So, I'm not quitting.

My teacher's father would pass away and she would be gone for two weeks during which time I was pretty much on my own. And I'll tell you now that I got through it all just fine.

I did get some help the very next day. With five minutes to go before the first period bell, and no substitute, I had given up on anyone showing. But then I heard a key in the door and eagerly went to open it. This caused a problem. "I can't get the key out," the sub said. "Close the door." I did so; he got the key out and opened the door. "My name is Jaime Ortega." he said. "Sorry I'm late; there was traffic. Is there a seating chart?"

I handed him the chart. "Yes," he said like a doctor looking at an X-ray. "I know a lot of these kids," he told me. "Some of them I know from when I subbed at the elementary schools. And I know some of their brothers and sisters." He was a former student teacher of Tina's as well, and had subbed in the school many times. He was also from the area. "I'm hungry," he said. "I worked out this morning and didn't get a chance to eat breakfast. We get a break after first period?" he asked.

For the next two weeks our routine began with Jaime telling me he was hungry and taking attendance. I would check in homework. After that I would do the warm-up problems, go over yesterday's homework, and start the day's lesson. Jaime would help answer students' questions and work with students who had trouble focusing.

That first day, we continued to work with exponents—this time presenting the rules for multiplying and dividing powers of the same base, like $5^6/5^2 = 5^4$. Which actually should have been the lesson for the day before so as to better explain negative and zero exponents.

After first period would come a break, and Jaime's daily routine was to go to the teacher's lounge and eat whatever snack he happened to have on hand.

Second period was the discovery-based algebra class. I checked in homework, Jaime would go over the answers to the homework; then I would start the day's lesson. It was about functions and graphing. From the second week of class they were "finding the pattern" of growth in a variety of problems. For example, a person buys a tree seedling that is 10 inches tall and plants it. It then grows at a rate of 2 inches per month. Students are asked how tall the tree is after 2 months, 5 months, and finally x months. They are supposed to figure out that 10 inches is the starting point, so they will end up with the equation $y = 2x + 10$ which they then graph and notice that the starting point of 10 crosses the y axis.

CPM wants students, to "make connections" between the patterns of growth, the equation for such pattern, table of values, graph of the equation and word description of the patterns. After all this, the book finally provides a definition of y-intercept and slope and the standard form of the equation of the line ($y = mx + b$). And then students get to "put it all together". Except for figuring out slope from any two points; that was still three chapters away.

The authors believe that "simply memorizing what to do in a specific situation without an understanding of the reasons why the method works too often leads to quick forgetting and no real long-term learning." To rectify this sad situation, students are "asked to solve problems designed to develop the method." In other words, teaching procedures directly never leads to understanding—o r connections.

Rudy was a very bright student, and despite earlier confusions in which he referred to 2x + 5 as "paste," he was definitely the "go to" person within his group of four. His group liked me because of my unusual habit of answering questions. He asked me about a problem that asked for the equation of the graph of a line. "Do you know what the y-intercept is?" I asked.

"Where it crosses the y axis?" he asked.

"Yes. Do you know the equation y = mx + b?" I asked pointing to the equation I had written on the board earlier. He nodded. "So what is the y-intercept in that equation?"

I reminded him of his past "connections": How to plug in the intercept value for 'b', how to compute slope, how 'm' means 'slope' in that equation. His eyes got big and he began speaking in Spanish to his group. Had Tina been there she would have told me I was giving them the answer. And I was—sort of. The problem-based "connections" approach wasn't leading to "aha" moments for many students. I noticed Jaime giving similarly explicit answers to the other groups.

Later in the day during 4th period prep, I went to the teacher's lounge. Jaime was there, eating salsa chips that he sprinkled with Tapatio sauce.

"I thought you had left for the day," I said.

"I was hungry," he said. "Chip?"

"Too hot for me," I said.

He appeared to be thinking about something. "You know, I am tutoring some students in calculus at the high school here and they use the CPM book for that. I don't think it's very good. You have to be really smart to see what's going on."

"I'm with you."

"Any word on when Tina is coming back?" he asked. I said no one knew anything. In the meantime, though, it was nice to have somebody with me on the raft—someone who was not afraid to say the sun was not rising in the south.

8. The Barber and the Grandpa

One thing I imagined I would do as a student teacher would be to ask my students "Who can show this problem who's boss?" whenever I wrote a problem on the board. Students would of course rise to the occasion. Nothing remotely like that ever happened. What did happen varied, but one event stands out. It occurred the Friday of the second week that my teacher was gone.

We had just covered squares, square roots and irrational numbers. I had shown with the aid of a spreadsheet how we can close in on the square root of 2 through successive approximations. I tried to get across that some numbers can be expressed as a ratio of two integers and others cannot. That idea remained an abstraction. My students could say that square root of 5 is irrational because it isn't a perfect square, but that was it. (I myself didn't really grasp this until high school.)

I ended the topic in my honors pre-algebra class saying "I know that the concept of rational and irrational numbers is abstract, but it's a foundation of what we call the 'real numbers'. If you study mathematics, you will see this concept many times."

To which a chubby, jovial boy named Arturo replied: "I'm going to be a barber."

I didn't know what to say, so I said nothing. This was at the end of the second week, as I say. A long week. On Monday of that week I announced the news to the class that their teacher's father passed away over the weekend. I said she would be gone all week, but back next Monday. This was also parent teacher conference week. Our classes were shortened, so students only had half a day. The afternoon was devoted to the conferences. "I will be available to meet with your parents, and Mr. Ortega will be with me as well."

In all my classes, I circulated a sympathy card in which I had placed an accordion style insert of extra paper so everyone could sign. Marta in first period told me "That's too bad her father died. I lost my grandpa last year; I was very sad."

"I'm sorry to hear that," I said.

Rebecca, who sat next to Marta said "My grandpa died too. A few months ago."

"I'm sorry," I said. She nodded.

The first person to talk to me at the parent teacher conference was a distinguished looking man, about my age. He was the grandfather of a girl named Gabby in my honors pre-algebra class. "I am the grandpa," he said, and shook my hand. Gabby had her two year old brother in her arms. "This is Louie," she said.

I explained to the grandpa the situation with my teacher but that I could talk to him about his daughter. He spoke in Spanish from then on, and Jaime translated. He wanted to know how Gabby was doing. I looked up her class grade; she was doing fine. "Does she talk a lot?" he asked. I told him that the people she sat with were often noisy and she sometimes joined in. All in all, she was fine. In fact, she was a delight. He looked stern and said something to her in Spanish and she nodded. He thanked me, shook my hand, and she said "Goodbye, Mr. Finn!"

The conferences were set up in the cafeteria. We were stationed at long tables with name cards. Families, often with little ones in tow, roamed around the room as if it were a train station. Not many parents came by. Those who did were concerned about their child's grades. I told them I would alert Mrs. Stevens when she returned, but also said that Mrs. Stevens and I were always available during lunch to tutor. And on Wednesdays after school, Mrs. Stevens also tutors—there are late buses. Very few students took advantage of this.

After I had talked with Gabby's grandfather Jaime told me "A lot of times the grandparents or other relatives are taking care of the kids."

"The parents are working in the fields?" I asked.

"Many of them, yes, but you have some kids who have a father in jail and sometimes the mother leaves. So they live with a relative or a guardian."

By the end of the week, I had found out that Jaime's parents worked in the fields, that he lived with them, that he had been a substitute for two years and there were few openings in the schools. He worked weekends at a pizzeria. In the summers, he helped his parents in the fields. And on Friday, it was his birthday.

"Happy birthday, Jaime!"

"Yes. I woke up today and looked in the mirror and said 'today I'm thirty'."

"Pretty young by my standards," I said.

"Yeah, maybe. For Mexicans it is not. In Mexico, if you're 30, you're married with kids."

The room was getting quieter; there were fewer families on that last day.

"I think people take a lot for granted here," he said. "A lot of kids think that because they're not in Mexico it's going to be easier. I tell the students, it's about choices. You can work hard and try to do something with your life, or you can work in the fields all your life."

"Or you can be a barber," I said.

"Yes," he said and smiled. "You can be a barber." He looked at his watch. "Well, there's not much going on, so I'm going to get something to eat."

"I guess I won't see you anymore after this," I said. "I enjoyed working with you."

"Same here. I sub a lot here so you'll see me."

"I hope so," I said and we shook hands.

I left shortly after. On the drive home I passed by many strawberry fields and wondered which ones Jaime had worked in, or my students' parents. When there were no more strawberry fields to see, I started wondering what it would be like to have Tina back in the classroom once again.

9. Not Quite Like Old Times

On the Monday that Tina returned, I was preparing for first period. I was always nervous before the first period bell rang, like an actor hearing the audience as he waits for the curtain to rise and the play to begin. I was particularly nervous today because there was a possibility that Tina might not show.

I heard the usual sounds outside my classroom—kids playing handball against the giant wall of the gym building that stood across from our module. As the school buses arrived, more kids poured into the courtyard. The lonely sound of handball was replaced with what sounded like 10,000 students milling around. I was relieved finally to hear the familiar clacking of high heels on the walkway to the classroom and the sound of the key in the door.

"Welcome back," I said.

"Thank you," Tina said, and headed for her desk. She saw the group sympathy card from the class and one from me and put both in her purse.

A math teacher came into the room, hugged Tina and gave her a small gift. The two chatted for a minute and then the teacher left. In the remaining few minutes, Tina tried to catch up on what was going on. "What are we doing in pre-algebra?" She looked tired and tense.

"Just starting Pythagorean Theorem," I said

"And algebra?"

"Multiplying binomials using generic rectangles."

She suddenly brightened and said "Don't you just love generic rectangles? It's such a great way to teach how to multiply binomials." She went on about how much she liked the approach, and how CPM would connect it to factoring later. I was glad to see her perk up and talk excitedly about CPM, even if I didn't care for it.

Generic rectangles are a way to represent the multiplication of two binomials as the area of a rectangle. So x + 5 multiplied by y + 10, can be represented as a rectangle with those binomials as the lengths of the sides:

	y	10
x	xy	10x
5	5y	50

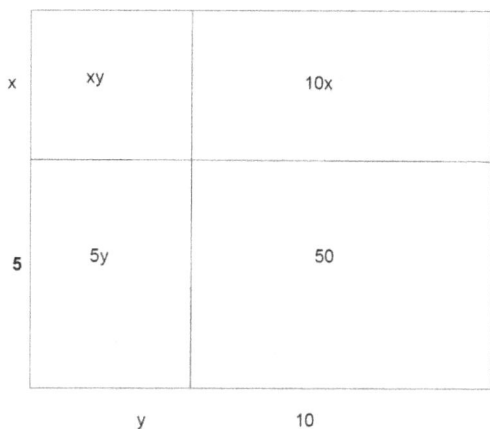

The students had experience computing areas within rectangles; now they could represent rectangles as above. They compute the areas of the four smaller rectangles, and add up the results to get xy + 5y + 10x + 50. The challenge I faced was not so much in teaching the students how to do it, but how to keep them from calculating (x+5)(y+10) using the procedure known as FOIL, which they had learned in 7th grade pre-algebra.

In fact, later that day in algebra class the issue of FOIL—the procedure for multiplying two binomials via 'First', 'Outer', 'Inner' and 'Last'—came up in a rather surprising way. I had been in charge of the class. I became concerned when I saw that Tina suddenly looked very tired; she sat down and put her head on the desk. I watched her out of the corner of my eye while I continued working with the students on the generic rectangle problems. One girl, Samantha said "We learned how to do this last year using this method. Why can't we use it?" She held up her notebook. I began to answer. "Yes, that's FOIL, but..."

Tina suddenly stood up and shouted: "No! Don't let them do it!" She then addressed the class: "You are NOT to use the FOIL method yet. I know some of you know it, but you have to understand what it is you're doing first."

She was fiercely loyal to the philosophy of the CPM authors who believed that the connection between binomial multiplication and area representation provided the understanding that students need. The authors were also fiercely loyal to the theory that teaching a procedure will frequently skimp on understanding. Well that's fine, but just use it to introduce the topic, then teach them how it's done algebraically. Do you really have to spend two weeks working with generic rectangles to instill "connections", "understanding" and other educational trendy ideas in order to teach them something they've already learned in pre-algebra?

In fact, the pre-algebra text had a pretty good explanation for how to multiply binomials prior to bringing in FOIL as a shortcut. They start with the distributive property: $a(b + 3) = ab + 3a$. Then if "a" is replaced by a binomial such as $b + 4$, $a(b+3)$ becomes $(b + 4)(b + 3)$. Substituting in the original equation, you get $(b + 4)b + 3(b+4)$, or $b^2 + 4b + 3b + 12$, which is $b^2 + 7b + 12$.

That's another thing that made me wonder. How much of this discovery and connection that CPM is bragging about is because of pre-algebra courses that students have had?

That aside, Tina was happy to see her students again, and they were happy to see her which I was glad to see. She continued on as if nothing had happened. Especially in the pre-algebra class. In that class, I had explained the Pythagorean Theorem, but I was a bit too slow for her, and forgot to give the formula, so she jumped in, just like the old days.

"Let me interrupt and point something out about the Pythagorean Theorem," she said, "because it wasn't clear." These last two words she said looking at me out of the sides of her eyes. I knew she wasn't pleased. She summarized it as "It's $a^2 + b^2 = c^2$. Can you say that?" The class repeated it. She then showed them the trick for how to identify the hypotenuse which is the "c" side in the equation: pretend the triangle is a bow, and the arrow goes where the little right angle sign is. The arrow is pointing at the third side: the hypotenuse. Funny how she could be so explicit in the pre-algebra class and yet buy in to CPM's philosophy in the algebra class, I remember thinking.

But it was good to see her again, and I told her so at the end of the day. "Good to see you too, Huck," she said. Something didn't feel right, though. It felt like we both were

trying to get back to the way things were before she left—
and that we both knew that would never happen.

10. Melting Ice, Feathers on Birds and Simplifying the Answer

In the course of my student teaching, I came across many questions, some of which had complicated answers, and others, simple explanations. One seemingly benign question plagued me for much of the school year. It presented itself at the beginning of the day after I parked my car and began my walk to the school. The roofs of the modular buildings would be dripping water onto the walkways beneath them and I would wonder every day where the water was coming from.

I had many explanations for the dripping water, most of which were quite complicated—lawn watering that went awry, and other theories. I found the answer about the last week of my time at the school. As I was heading toward school from the parking lot I saw a young woman who taught in the classroom next door to me. I asked her where the water was coming from. "Oh, that's from the ice melting on the roofs. At night it gets below freezing and the moisture on the buildings freezes. In the morning sun, the ice melts."

While the explanation was simple, I felt as if I had been told how the universe was created. It also hinted of a sadness for reasons I've never been able to explain. The teacher who told me the secret of the ice melting, taught a class in basic English for those students who spoke only Spanish. During the 4th period of the day, which was the "prep" period for Tina and me, I would hear her students next door reciting their sentences in English. One sentence which I heard quite often was "The feathers are on the bird", said in unison by thirty or so voices, all pronouncing "the" like "thee".

The school had what was called a "modified block", in which math classes were held every day rather than every other day. After my three 83 minute periods, there was lunch, and then came the 4th period.

Fourth period was like an after-hours time. If Tina and I weren't preparing for an activity the next day, Tina would go about her business elsewhere, and I would be left alone in the classroom. I would sometimes work on the next day's lesson, sometimes correct tests, and other times prepare upcoming tests or quizzes. On other days, Tina would be in the room with me.

It was a neutral time. She rarely talked about anything I happened to do wrong during the day. If she did, it was quick and then on to other things. Sometimes she would talk about her son and the problems he was having in high school. There were times when we worked on something together. We would find things to laugh about—hysterically like two teenagers—in the giddiness that sometimes comes when two people are punchy from exhaustion. And other times, we would both work in silence. Sometimes her head would be on her desk.

The day I found out the explanation for the dripping water, I was working during 4th period, finishing the upcoming test and ran the questions by her to get her approval. She looked it over. "Why do you say multiply for number 6? She asked. The problem was: 5^4.

"I was avoiding saying 'simplify', since that word confuses a lot of students. It's already simplified. I want them to multiply it out."

"Why not say 'Evaluate'?"

"Yeah, that'll work," I said.

"The rest of the test looks fine," she said.

I was intrigued, though, so I went on. "I think 'simplify' is one of those words that really confuses students. I've seen problems in the book that say 'Simplify 5 x 5 x 5 x 5'. That could either mean 625 or 5^4. If they want 5^4, they should say 'Write in exponential form'."

Tina wasn't too interested in this conversation about "simplify", however fascinating it was to me. I went back to the computer to revise the question.

"I have a question for you," she said. Ah, she is interested in the topic, I thought.

"OK," I said.

"Didn't you tell me your father died not too long ago?" Not quite what I was expecting.

"Yes, three years ago," I said.

"How long did it take to get over it?"

A difficult question. I answered it as simply as I could. "I don't think you ever get over the sense of loss," I said. "But you do get over the sadness. And you start remembering good things about him. But it took a while. A few months maybe. But my father was a lot older than yours was, so it wasn't unexpected."

She said nothing, so I went on. "I remember that when he was alive, if I was getting ready to visit my parents, or call them on the phone I would have an imaginary conversation – imagining what he would say; things that might make me angry, and how I would deal with it."

She was still silent, evidently wondering where I was going with this.

"But after he died, I would sit in a chair in the living room after I came home from work and would wonder why it seemed so quiet in the room. It was unearthly, it was so quiet. And I realized one day, it was that his voice was no longer in my head. So I had to get used to the silence."

I realized I probably shouldn't have said all that. Tina looked pretty sad. Sometimes it's best to just leave things at "the feathers are on the bird" and "the water comes from the melting ice".

There were five more minutes until the bell and the principal was strict about teachers not leaving before then. It seemed like I should say something else, but I could see we both thought it best if I left, so I did.

"See you tomorrow," I said. I made it to my car and out of the parking lot before the dismissal bell all while saying a silent prayer for her.

11. I Say Goodbye and Head Out for the Territories

Tina's moods were an up-down affair. On one day she could be very pleasant; but the next day she would be super-critical and negative. I tried working around her moods, but it was difficult. I knew her moods were due to her father passing away. And in all fairness, her advice to me—though annoying—was generally pretty good.

"When you ask a question, you have to give them time to answer it," she would tell me. "Otherwise, they know you'll just tell them the answer, and you'll never get a response from them." This is valuable advice—wait time is critical.

And in the pre-algebra classes I taught, we used a direct instruction technique which entailed asking questions—and waiting for answers. Critics of the direct instruction method will tell you that it is all about kids sitting in rows, facing front, while the teacher lectures and provides instruction by rote. Well, first of all, our students were seated in groups of four, though I would have preferred rows. Secondly, I tried to engage the students by asking questions—not just "lecturing". Third, with the exception of the Pythagorean Theorem, there was nothing rote about how we taught.

In the discovery-based algebra class, students were expected to work in groups and come up with an answer, much of the time without the benefit of prior instruction. The method they were to have discovered would often be revealed in the next section. No one ever seemed to catch on to this. Tina's admonishment to me in that class was similar: "You're telling them the answers; let them figure it out themselves."

I tried to follow this advice while they were learning to solve systems of equations with two variables. They had learned about eliminating variables; that is, given two equations like $x + y = 6$ and $5x - 2y = 8$, you can multiply the first equation by 2 to obtain $2x + 2y = 12$. Adding the two equations result in the $2y$'s dropping out. But suddenly the book presented these two equations: $4x - 3y = 1$ and $3x - 4y = -1$. The students were stumped. Even the brightest student, Jorge, didn't know what to do. He asked for help.

Tina was out of the room. That was the other thing. While she criticized me a lot, she also gave me time alone with the class.

I wanted to stay true to what Tina wanted, so I did the following, knowing that Jorge was quite bright and would catch on to the hint I was about to give him. "Remember when you were learning how to add fractions, what you did when you added $\frac{1}{3}$ and $\frac{1}{4}$?"

"Yeah, what about it?"

"What did you do?"

"Found a common denominator," he said, clearly not getting where I was going.

"Which was what?"

"Twelve."

"How did you do that?

"Multiplied 3 by 4."

"OK," I said. "Look at your equation. You have a 3y up here and a 4y in the second equation."

"OH!" he said "You multiply the top one by 3 and the bottom one by 4."

Now I imagine Tina might have criticized that, but I felt I wasn't giving it away and was using proper scaffolding techniques. I tried the same technique with other students, and they eventually saw where I was going.

I guess when push comes to shove, although I dislike the CPM program, I've seen worse. It covers what needs to be covered. This doesn't let CPM off the hook. It could be done a lot more efficiently with a lot more practice problems. Going back to first principles like generic rectangles and guess and check leaves a lot to be desired and takes credit for the hard work teachers do in the pre-algebra courses.

I also think that Tina gave more direct instruction in that class than she was willing to admit. There were plenty of times when kids weren't getting it that she would stop everything and offer instruction at the front board. All in all, I think Tina is a good teacher and relies more on direct instruction than she might think. She often advised me on things that sounded an awful lot like things I would say.

She gave me a nice farewell, and in the last week or so, told the kids that I would be leaving soon. They seemed sad about this. A girl named Gabriela in the algebra class asked me if I was leaving because I was moving away. This seemed likely to her since students were used to seeing their friends leave because of "moving". Some parents moved to another part of town, or out of town, sometimes because of rent, or losing homes, or finding new jobs. I explained that I was finishing up the program; my teaching was an assignment, just like her algebra course was an assignment. I don't know if she understood or if I remained just one more casualty of moving in her mind. In any event, the irony of her question caught me off guard.

"Why are your eyes watering, Mr. Finn?" Gabriela asked.

I was ready for that. I was always ready for student questions. "Oh, you know I have terrible allergies and they're mowing the lawns today." They were in fact. "Whenever that happens, my eyes just itch and burn." She seemed satisfied with the answer, but I don't know if she believed me.

Although sometimes I couldn't wait for student teaching to be done, I miss working with Tina and I think of my students often, with great fondness. I tried to teach them as if what I was teaching mattered. I have an idea of who it mattered to and who it didn't. For most, though, I simply couldn't tell.

Part II: Out in the Territories, More or Less

12. My Triumphant Return: Trying to Find a Cure for the Flu and the Common Core

I have decided to take up my correspondence once more with Miss Katharine about my experiences rafting along the ideological, political, and cultural river known as math education. Some of you may dimly recall that about a year ago I wrote some letters using the name Huck Finn, describing my experiences as a student teacher. I am not yet teaching full time, but I am subbing, so you'll have to be satisfied with that for now. I've also abandoned the raft for a canoe for easier maneuvering.

I chose the name Huck Finn for two reasons: 1) I am looking to get hired as a teacher and because my opinions may not be popular amongst those on the other side of the river, my real name is best not told; and 2) the sheer poetry of it. My experiences have for the most part been very positive and instructional. I missed my old class and my teacher from my student teaching days and thought that no one would ever take their place in my heart. For the most part it's true, but I do recall on one of my early subbing jobs at a middle school, the first student entered boldly into the classroom, overjoyed at the prospect of having a sub instead of the real teacher and cheerily announced: "Hi! My name is Lupita, but you can call me 'sexy'. " I replied "Hi! My name is Mr. Finn, and I will call you Lupita." I was amazed at the reflexive and automatic nature of this response.

It is the heart of flu season and my greetings from students have gone from "call me 'sexy'" to "Can I go to the nurse's office; I feel like I'm going to throw up." In the course of my work, things don't get any more complicated than that. I usually don't find myself in the thick of political or ideological battles concerning math education with anyone. But the other day, something did happen to make me aware once again of my trip down the ideological river. I was subbing in a high school geometry class, and preparing to put up a quiz on the projector when I noticed on the table at the front of the class some extras of handouts. One handout was titled "Standards for Mathematical Practice" which consisted of the eight standards that relate to how math should be taught, per the Common Core Math Standards, which have been adopted in 46 states. I picked it up, and realized it had been handed out to students as a list of things for them to do in the course of "doing math".

With the handout in my hand, I was suddenly surprised by the sound of my own voice. "I really hate to see stuff like this," I said. The students looked up. "It means nothing to you, I know, but it comes from something called Common Core which will go into effect next year. It is going to result in math being taught in certain ways. In fact, it's where your teacher is today, at a conference on how to teach the Common Core standards, learning the new ways she has to teach you math." Fortunately, the piqued interest of the students had now turned into blank stares. Their years of acquired expertise had taught them how to tune out teacher soliloquies.

The Standards of Mathematical Practice that got me so excited are the following:

1. *Make sense of problem solving and persevere in solving them*

2. *Reason abstractly and quantitatively*

3. *Construct viable arguments and critique the reasoning of others*

4. *Model with mathematics*

5. *Use appropriate tools strategically*

6. *Attend to precision*

7. Look for and make use of structure

8. Look for and express regularity in repeated reasoning

Taken at face value, they are as seemingly benign as when the company you're working for says that the latest reorganization will not change anything and your life will continue as it always has. "Make sense of problem solving and persevere in solving them" seems like good advice, and the way I interpret it is to help students tackle problems by giving them the tools, instruction, guidance and practice to do so. One of the many ways it is being interpreted, however, is to require students to find multiple ways to solve problems. (I say this because I noticed in another classroom a poster bearing the logo of "Common Core" at the top, showing multiple ways of doing particular types of algebra problems.)

There's nothing wrong with finding multiple ways of solving problems. But in early grades, students find it more than a little frustrating to be told to find three ways of adding 17 + 69. Putting students in the position of satisfying the teacher only if they find multiple ways of doing it is a recipe for 1) disaster and 2) rote learning, the bugaboo of the purveyors of "find more than one way to solve it".

Now it's true that during my recent assignment, I went over a proof of a problem they had in their homework and, immediately after going through the proof the teacher had provided, asked "There's another way to do this proof; can anyone see how?" I'm in a habit of doing that in high school math classes, and I suppose that's good, because if it comes to the point that I actually teach full time and a strict principal decides to make sure I'm following the Standards for Math Practice, he or she would likely conclude that I'm doing it.

What I will not be doing is *requiring* students to do things in multiple ways. I'm a bit stubborn on this point, but I tend to believe if a student can do a proof, or solve a problem and do so correctly by applying prior knowledge, then he or she doesn't also have to do twenty five fingertip pushups.

All in all, teaching high school is a bit more straightforward than the lower grades. By that, I mean that the span of content to be covered is such that there isn't room for discovery- and inquiry-based group work and "read my mind—what answer am I looking for" type of catechisms. I hope that Common Core doesn't bring that about. If it does, you can look for me along with flu-stricken students in the nurse's office, trying to keep the nausea at bay and figuring out what my career options might be.

13. The Trombonist in the Classroom and a Paper I'll Probably Never Write

In my substitute teaching I often see students who remind me of my former students during my student teaching days. During a recent sub assignment, after I gave the students their assignment, one chubby boy asked me if they could work with partners. "Yes," I replied.

"Ooh! 'Partner' is a dirty word!" he said.

I immediately thought of Angel, a chubby 7th grade boy in my honors pre-algebra. While the purveyors of reform math and trendy methods for teaching math always talk about getting students to use their innate abilities in math to make "connections" between aspects of math, they fail to account for the innate genius that middle schoolers possess in making sexual connections. I saw this while finishing a lesson on negative and zero exponents for the honors class. I had started to say "I realize that these concepts are confusing and it's hard to wrap your head around them, but…"

That was as far as I got. I heard a tittering and suppressed giggles, so I went on to the next subject. Whenever I would turn to write something on the white board the giggling would get louder. The giggling continued, and at one point I saw that Angel had his head down on the desk, convulsing so hard with laughter that I thought he was crying. I stared at him to see what was wrong. He realized I was looking at him, and he stopped.

I didn't figure it out until later when I was driving home, which is when many things tend to hit me. "Oh, 'Wrap your head around it!' I get it!" I said to myself.

I can't blame them for laughing. I recall something similar happening during a rehearsal when I was in the Michigan Marching Band. We were rehearsing the week's pre-game and half-time show numbers indoors. The legendary William D. Revelli was conducting; he was a perfectionist and something was not right among the trumpets. He had them play a passage over and over while the rest of the Band sat, not playing. He suddenly addressed us non-players and said "Boys, when you're not playing your instruments you can still be practicing by fingering your parts." I remember the great difficulty we had stifling our laughter. In fact, to this day, if I need to smile for a photo, I will recall that incident—to good effect, apparently since people generally remark on the nice smile I have in the photo.

Angel, the perpetrator of many antics in the honors pre-algebra class, was in the school's band. I found out one day he played the trombone, which delighted me, because my experience with trombone players through the years is that they tend to be troublemakers. I cling to a theory that cannot be proven, that there are personality traits associated with the instruments people choose, and so I have found that trombone players are a loud, jolly, mischievous bunch who mean well. They tend to contribute humor into the class, as well as a sober grounding in reality.

I recall one day when I was teaching a unit on exponentials. This was during the two-week absence of my supervising teacher, Tina, after her father had passed away. I was going over common errors that students make when doing problems like $7^2 \times 7^3$—like writing the answer as 49^5, and representing $7^6/7^4$ as 1^2. "What is $7^6/7^4$?" I asked. They answered "7 squared". "That's right," I said. "So why did so many of you on the last test answer the question as 1^2?" I asked, writing that on the board.

A few minutes later, I was going around the class monitoring their in-class assignment. I saw that Angel had made this same mistake. "Angel, why did you do that? Didn't you hear me explain that the base remains the same when you divide powers?"

I was standing behind him and he tilted his chubby head back and looked up at me. "But that's what you had written on the board," he said, and pointed to the example of the mistake that I had neglected to erase.

I relayed this tale to Tina sometime later after she returned, in one of our moments of catching her up to what had gone on in class. "Not a good idea to show students how to do something wrong," she said. "You'll have kids like Angel who only half listen and they'll think the wrong way you wrote on the board is the right way." I've half-followed this advice. I do think it's sometimes valuable to highlight something students are doing wrong, if many of them are doing it. But if I illustrate the mistake on the board, I erase it as soon as I'm done talking about it.

I don't know that Angel should have been placed in the honors class, but then again, there were so many factors that affected the students' performance. While some people I've met would be quick to classify him as being of "low cognitive ability," I find it hard to make such identifications quickly. I recall two girls who were doing poorly in the pre-algebra class—one was in honors, and the other in the non-honors class. The teacher held conferences with the girls' mothers, both of whom worked in the strawberry fields. There were problems at home that got sorted out and the girls' performance soared—I couldn't believe they were the same people. I knew it was a respite—home life issues rarely are permanently solved.

Towards the end of the semester, it was nearing the date of the school's band concert. A few minutes before class started, Angel approached me for help tying a necktie that he had to wear for the concert. "I'll give it a try; I'm not used to tying it on someone else." I got it to loop correctly and slide up and down, though the narrow part was way too short. "The length doesn't look right, I'm afraid," I said.

"Oh, this'll be fine," he said, delighted. He removed the tie keeping the knot intact.

"Ask your father if he can adjust it for you," I said.

"My father doesn't wear ties," he said. I suppose I knew that and was embarrassed at having said it, but in thinking about it later I realized Angel was just stating a fact, not making a statement.

And so let me end with both a fact and a statement: I know I lack experience and am hopelessly naïve (that's the fact) but I hope I don't end up hung up on IQ's and other measures of cognitive ability. (That's the statement). Even bright (high IQ) kids have to get help at home to succeed. Maybe poor prior math instruction and disruptive family life play a role. Maybe cultural factors play a role. Maybe the cognitive power of making sexual connections is a surrogate for IQ. Maybe I'll write a paper about it someday. Believe me, I've read far worse.

14. I Sneak in the Back Window and Teach How to Attend to Precision

Dedicated readers of my letters may recall my reaction to the Common Core's "Standards for Mathematical Practice" (SMPs). I continue to see these SMPs posted in various math classrooms where I sub, though I try to ignore them as much as I can as I continue to drift down the ideological divide, otherwise known as math teaching.

The SMPs came to my attention again recently while subbing at a middle school. For me the day started as sub assignments usually do: by reading the teacher's directions to the sub. That day, the teacher wanted me to administer a "district assessment" for each of her classes and then have them start on their homework.

I glanced at the assessments and saw some mention of "Common Core" on the front of the answer sheet. Neither the answer sheets nor the instructions that I had to read aloud gave the reason for these tests, other than that the students would be evaluated as to how they solved and analyzed problems. I expected that students would ask me if their tests figured into their grades, so I called the front office before first period class and asked.

"I'm not sure," said the person in the front office. Pause. "Well, let me think." Pause. "No. I'm pretty sure they don't figure into their grade."

"Thanks," I said. "I will tell the class what you told me." Accountability did not seem to faze her and she said "OK." (I since found out that, yes, it does affect their grades in class, and their placements next year, and is a performance task like what they will see with the Common Core when it goes into effect—next year.)

The test had four problems which students had to answer by showing their work directly on the exam and "explaining their answers." Before my first period class started the test, one boy raised his hand. "Our teacher always tells us to 'Attend to precision,'" he said, pointing to that particular SMP which was posted on the wall behind me. "Could you say it please?" The class looked at me expectantly.

"Attend to precision."

"Thank you," he said.

"You know, I don't even know what that means."

"It's one of the things she has on the wall," the student said.

"Yes, I know. But it's just so vague. Here's what it means to me. It means use the right math vocabulary, and show your work. So for this test, if you just write a number down for your answer, it won't be enough. You have to write what you did so someone else can follow it." This seemed to satisfy the class, which is to say that probably not one person focused on what I said.

The questions had to do with discounts and percentages, which they seemed comfortable with. One question asked them to say whether an item was marked down 25% four times in a row, explain whether or not the final price after the four discounts would be $0, and provide the reason for their answer. From what I could see when I glanced at the tests while collecting them, most students didn't get the answer right, but they had no problem putting down an explanation.

The situation was quite different with the eighth grade algebra classes, however. Their test only had two questions. One involved an L shaped figure with dimensions like x+15 on one side; i.e., no simple numbers were used. The question asked for the students to write an expression for the area of the figure. This involved splitting the figure into two rectangles, figuring out what various missing dimensions were, and then writing the area as the sum of the area of two rectangles—something that would amount to an algebraic expression.

After a few minutes of quiet, there was suddenly a flurry of indignant questions: "Do they want us to calculate the actual area? Like a number?" "What does 'expression' mean? Is that like an equation?" I was struck by the difference between the seventh graders who simply wrote down their process, and the eighth graders who were confused by what was required of them. One student put it into perspective for me. "What do they mean 'explain your reasoning'? I just do it." I took this to mean that all year they've learned how to express ideas algebraically, with showing their work being sufficient explanation. Given that that's what they thought they were doing to begin with, requiring them to "explain their reasoning" made no sense.

Of course, the students would not know that there are people who view those who can't "explain their reasoning" (however correctly they solve a complex problem) to be doing "rote work" and lacking "understanding." But it seems to me that if we really want students to do such explaining, then we should tell them how. Simply telling students to "explain your reasoning and attend to precision" is not likely to accomplish much. Knowing how to explain something precisely doesn't come automatically with understanding. And students are not likely to pick it up by themselves working in groups and the like.

As it was, I had a class full of eighth graders at a near-riot level. I stood in front of the class and said, "OK, I'm going to show you something." They were still talking. "Please listen," I said and then added "Actually, please don't listen because I'm probably going to get fired and/or shot for doing this." The class immediately got quiet.

I drew a square on the board and labeled two of the sides with an "x". I said "This is a square with sides of x units. How do we express the area?" The class said "X squared."

"Can you explain why?"

Someone said "Area is length times width."

"That's all I'm going to tell you," I said and erased the figure.

I'm still waiting for someone to knock on my front door and put the handcuffs on me.

15. The Diversity of My Experience

The other day I was at a middle school in a 7th grade math classroom. The class had taken the Common Core district assessment that I described and they now had to do their homework, which was difficult given that there are only 28 school days left to the school year, plus I'm only a sub, and the weather is nice.

In the midst of my telling multiple students to get in their seats and do their work, one boy confronted me and said, "You're just picking on me because I'm Mexican."

"No, I'm not," I said.

He pointed to some students talking and said, "You didn't tell them to be quiet." True. I told some students on the other side of the room. I didn't continue the argument but told him to sit down and do his work.

In my travels down the cultural divide, I have had to adjust from my student teaching days when all my classes were 100 percent Mexican students to my present situation in which my classes now are mixed and Mexicans are in the minority. Rules are in place in all schools against prejudicial and hate language—to the extent that middle-schoolers will sometimes try to blame someone for saying something that is racist in the hopes of getting that person in trouble. I have heard comments muttered about Mexicans, but nothing loud enough for me to "write anyone up".

Since the incident with the Mexican boy, I've been thinking back to my student teaching days in the heart of strawberry country. My drive was a tiring one—an hour and fifteen minutes each way. The majority of the drive was along Highway 101, one of the major north-south highways of California. It took me past farms that grew lettuce, broccoli, tomatoes, apples, avocados, alfalfa, and strawberries. The strawberry fields became more plentiful the closer I got to my destination.

A long bridge that crossed over a dry riverbed and the county line marked my entry into the city where my school was. There used to be shrubbery in front of the school but it was removed because it became a place where students could hide weapons. I was told this on my first day. I was also told that there are two gangs in the area, and there are gang signs for each. Displaying a gang sign while at school is grounds for suspension. Drinking bottles and water bottles are banned because students were putting alcoholic beverages in them. This was discovered the year before I started, when a student passed out and had to be taken to a hospital. Students can be suspended for fighting, for using racial and sexual slurs, for continued violations of rules, and for talking back to a teacher.

During my 15 weeks there, I did not come into contact with these things any more than I come into contact with the ills of society in my everyday life. What I did see were the effects of one-parent families, students having to miss school so they could care for their baby brothers or sisters on days both parents were working in the fields, or, in the case of a boy named Isaiah, missing school because of ongoing doctor appointments due to severe allergies. He was asthmatic, could not go near grass, and frequently experienced nausea and hives in addition to asthma attacks.

My supervising teacher, Tina, thought Isaiah's allergies were because of the pesticides sprayed on the fields. I sometimes smelled pesticides when the helicopters would spray the various fields. I never saw the helicopters, but Tina would see them sometimes. She told me one day about a helicopter she saw spraying a field and the clouds of gas that spilled into the neighborhoods where kids were walking to school. "It's no wonder we have so many kids with respiratory problems," she said.

bell!" My voice was loud and my hands were shaking. I only made it to 30 seconds after the bell before I told them to leave.

It's difficult to know what is going on with a student without knowing the background. Some might think that Miguel was of "low cognitive ability". But I recall two girls in my student teaching days who, after my supervising teacher, Tina, held a conference with their parents, began getting high scores on tests and participating in class—until family life deteriorated again. Another student from those days, Antonio, was very smart but decided to goof off and was in danger of failing. Tina held a conference with Antonio and his father who told him "If you fail this class, you will work with me in the fields this summer." Given such a choice Antonio probably realized what my daughter had learned: solving equations is a lot easier. He passed.

Then there are some who simply have given up. The reasons are often unclear—whether you're a teacher or a sub. I do believe that given the right math teaching from first grade on, if students put in the effort, they can get through algebra. Many think I'm a fool for thinking this way. Call me what you will.

School ends soon and I just had an interview for a math teaching position that looks promising, but that's happened before so I'm not going to make any predictions. In any case, I don't think I have that much more to say, so this will be my last letter to Miss Katharine. I thank her for allowing me to tell what I think needed to be told and thank the rest of you for reading them.

Your Pal,

Huck

E.W. Kemble.

Isaiah was a quiet boy who managed to get the highest scores on tests in the class if he didn't get too far behind in his homework. He didn't like to speak up in class and so I got into the habit of leaving him alone—I could see he knew the material. One day, shortly after Tina returned to class after her father passed away, I was teaching and hadn't noticed that Isaiah had his head down on his desk. Tina did, however, and she suddenly came over to him and asked if he were OK.

"Come on, Isaiah, I'll take you to the nurse's office. He stumbled to his feet and she grabbed hold of him to keep him from falling. I managed to continue through the day's lesson, assuring the students Isaiah would be fine. And he was. Fifteen minutes later, Tina and Isaiah came back to class; he just needed a dose from an inhaler.

Mercifully, Tina did not mention that I should have been more aware of what was going on, and should have seen that Isaiah was ill. She didn't need to. Since that time, I have made it a point to be more vigilant and have excused students who looked ill to go see the nurse.

My subbing assignments are no longer in the heart of strawberry country, though there are fields near some of the schools in which I sub. Also, as I said, my classes are now more "diverse," to use a term I've never liked because it's phony. For one thing, there are some who would say that my totally homogeneous Mexican classes were diverse because they were non-white. Also, the term reminds me of the Weekly Reader exercises from long ago that asked students to find four "friends" in a picture of a parade or something similar. The "friends" of course would be the distinctly foreign people. The "friend" who thought I was picking on him was using diversity and school policy against racist actions to his advantage. He wasn't very good at victimization and definitely needed improvement in math. I'm pulling for improvement in the latter.

16. I Guess I'm Just a Cheater

With all the scandals about teachers and administrators cheating to raise their students' scores on standardized tests, Miss Katharine was a bit concerned over my letter of a few weeks ago in which I described how I handled an assessment I had to administer. I told her she didn't have to worry: no one was out looking for me, and the only knock at my front door was from a young man who was selling magazine subscriptions that would serve some worthy cause which had benefitted him somehow and which slips my mind at the moment.

Recently I was back at the same school where I had administered that Common Core flavored exam and ran into one of the students from the 8th grade algebra class. They were the ones who had demanded to know what "explain your reasoning" meant.

"Oh, you're the guy who gave us that explanation for how to do the problem that no one could do," she said. Interesting that the explanation I gave took less than a minute but apparently was enough to get the point across; so much so that I've become somewhat of a legendary sub. Which perhaps suggests that explicit and direct instruction might not be the "rote learning" approach feared by otherwise sane and pleasant people.

The "understanding" and "connection" mantras are prevalent in the groupthink that makes up much of the education establishment's view of math education. It was at play big time with the algebra program I had to use when I was student teaching—a program called CPM algebra. I've mentioned it before and it seems appropriate to mention it again what with Common Core surfacing and being interpreted along the ideologies of reform math.

With CPM algebra, students were taught "slope" in a series of discovery lessons that spanned many weeks. They had to make "connections" between tables of values, and equations, how both described the patterns of growth, how the y-intercept value helped to draw a graph, how to draw a graph given an equation, and how to determine the slope and y-intercept when looking at a graph. All well and good, but the point-slope form of finding an equation was not presented initially; students were left in the dark for quite a while about how to determine the equation of a line given the coordinates of any two points on the line.

I kept to the script of the algebra text as best I could. This turned out to be disturbingly easy. You just went over the previous problems, gave a short intro for the topic of the day to get them going and then assigned the problems in the book for that day. They then worked on them in groups and Tina and I circulated to answer questions. I could see how if a teacher were lazy (which I hasten to say Tina, my supervising teacher, was not), they wouldn't have to do very much, and lesson plans were pretty much automatic. Tina bought into the program; she believed in it and worked hard to make it work. But it also seemed she was seeing what she wanted to see. There were times when, circulating around the classroom, she would say to me "They're getting it! They're making connections!" Yet, there were students who seemed quite confused and some of them knew that if they pushed me hard enough during my circulatory tour of the classroom, my hints (given while Tina was working with other students and out of earshot) would often tell them what they were supposed to discover. Maybe the connections she was seeing the students make were because of that.

The reason why CPM eschews procedures like the point-slope method of finding an equation is that it supposedly gets in the way of true understanding. I heard this recently from a teacher, in fact: "Kids buy into the slope formula, plug in numbers, do the calculations and yet they still do not understand what they are doing. They are simply memorizing yet another formula for some unknown reason."

I don't know. I just don't find slope all that terribly difficult to understand. Similar triangles and proportion seem to explain why the slope of a straight line is always going to be the same for any two points you pick. But people seem to think that if a kid is doing procedures without "complete and true understanding" he's doomed to a life of failure. It is as if the moment a student stops doing all the intermediate steps/algorithms and fails to make the appropriate connections each time, then he or she is using a trick or rote memorization to jump to the end result and not using understanding or strategies to solve something.

I recall one time when student teaching, talking to a fellow math teacher. This was during the time that Tina was gone for two weeks when her father passed away. The teacher was telling me about the math teaching philosophy. "Tina always says we can teach them how, but what's really important is that they understand 'why'". As she told me that, she looked to me as if she wanted me to say something. I sensed that underneath it all, she felt the same way I did— but was afraid of being disloyal.

I think of that hallway conversation often. I think of it when I see the posters for the Standards for Mathematical Practice on the walls of the various classrooms in which I substitute. They make me feel as if I'm back as a student teacher, trying to figure out the best way through a ridiculous program. And despite my strong beliefs about what I talk about here, I still feel like I'm cheating when I teach the way

I see fit, as if maybe 1) there's something wrong with me, or 2) I'm being disloyal. I've only met a few teachers who have told me they don't like the trends I've been describing in math education. They've usually been teaching for over 30 years and are about to retire.

17. Rules of Engagement and a Fond Farewell

I've been told that bribery is not a good way to motivate your kids or your students. I'm afraid I break that rule quite a bit. I broke that rule long before I started teaching, when my daughter was in middle school. Her pre-algebra class was learning two step equations and she was doing fairly well. I felt with some more practice she would become fluent. So I brought out my old algebra book and turned to a page with a bunch of the same type of problems and offered her the following deal: "There are ten problems here just like what you've been doing. If you do these ten problems I'll give you a quarter for each one you get right. So you can earn up to $2.50!"

At that time, her allowance was based on completion of certain chores. She did some quick reasoning and rejected my offer, explaining she could make that same amount by cleaning the toilets. Despite the warnings against bribing your kids, I saw the opportunity for a teaching moment and I told her "You're right. But I'll let you in on a secret. Solving these equations is a whole lot easier." She looked at me. "Think about it," I said. "Tell me what you decide."

The next day, she came up to me and asked "Is the deal with the equations still on?" I was thus started on a career of carefully controlled bribe-based motivations.

Along with other motivational and engagement techniques, I use bribery carefully—I don't overdo it. I will sometimes offer a quarter or fifty cents to the first student who can do a certain problem; it depends how much change I have in my pocket. Usually one student will respond, and I award the prize. In more advanced classes, like algebra 2, the prize is generally $1.00 and I require the student to explain their answer.

I find the most motivating prize (at least in the lower math classes) is dismissal two minutes early. As often happens, students begin to put their books away five minutes before class ends, and sometimes line up at the door. Rather than tell the students to get back in their seats, I will put a problem on the board pertinent to the day's work and tell them: "OK, here's the deal! Who wants to leave two minutes early?" The noise level of the classroom suddenly drops to that of a concert hall when the conductor raises his baton. "If anyone can do this problem, you all get to leave two minutes early."

Suddenly there is a flurry of activity. The smartest students are drafted into service by the others and attempt to solve the problem. I try to make it challenging enough that it takes some effort. If the first attempts are wrong, more students get into action—you would think I was offering $100. At last someone gets it and I dismiss them, telling them to please be quiet in the hallways or I will be fired or executed or face some other grisly end.

There has to be a critical mass for these techniques to work, however. I had a class a few weeks ago that the teacher told me to be firm with because "they tend to get squirrely" which roughly translated means "They will cause you to seriously question your decision to go into teaching."

To check what I was dealing with, I looked at the grade roster for the students in this class. About a quarter of the class had F's and another quarter had D's. This was the two-year algebra 1, and most of the students were ninth graders. Ninth graders are hard enough to deal with, but this was late April, so that those with F's had pretty much given up.

I tried to help one boy, Miguel, but he was far behind. He did one problem with a lot of guidance from me so much so that I realized he clearly was lacking some basic knowledge. With the second problem, he put his head down on the desk and said, "I just can't do it."

In the last few minutes of class, students started lining up at the door. I knew no motivation was going to work with this class so I yelled "Sit back down!" Miguel took that opportunity to bolt out the door. I lost it at that point. "OK, everyone in your seats; you will stay one minute after the